SpringerBriefs in Physics

More information about this series at http://www.springer.com/series/8902

Jijun Zhao · Lizhao Liu · Fen Li

Graphene Oxide: Physics and Applications

 Springer

Jijun Zhao
Lizhao Liu
Fen Li
Dalian University of Technology
Dalian
China

ISSN 2191-5423   ISSN 2191-5431   (electronic)
ISBN 978-3-662-44828-1   ISBN 978-3-662-44829-8   (eBook)
DOI 10.1007/978-3-662-44829-8

Library of Congress Control Number: 2014954227

Springer Heidelberg New York Dordrecht London

Printed on acid-free paper

Springer is part of Springer Science+Business Media (www.springer.com)

Preface

This book covers the current research on graphene oxide (GO), including its synthesis and characterization, structural modeling and fundamental physical properties, as well as potential applications, aiming to giving a comprehensive reference for scientists in the related fields or graduate students who want to know about GO.

Graphite oxide is a substance known for almost 150 years. The monolayer sheet of graphite oxide, namely graphene oxide, is not only an important raw material to synthesize graphene, but also shows excellent physical and chemical properties that might find important applications in electronics and optics. Meanwhile, existence of various oxygen-containing functional groups also enables GO promising prospects in energy and environmental science, as well as biotechnology. Due to numerous literatures in this fast-growing field, here we only highlight the most essential progresses instead of presenting an exhaustive bibliography. Eight chapters included in this book are summarized as follows.

Chapter 1 begins with the fabrication and reduction of GO. Three main methods to fabricate graphite oxide are described. Monolayer or multilayer GO sheets can be obtained from exfoliating graphite oxide. On the other hand, the production of graphene from reduction of GO is also introduced, focusing on chemical reduction of GO and its atomistic mechanism.

Chapter 2 illustrates the structural characterization of GO by means of both spectroscopic and microscopic approaches. In general, spectroscopic techniques (e.g., NMR, XPS, and FT-IR) can directly feature the atomic structures of GO, including the types of oxygenated functional groups and their distributions. On the other hand, microscopic techniques (e.g., TEM, STM, and AFM) can provide essential insights into the lattice atoms and topological defects.

Chapter 3 covers the theoretical modeling of GO structure, as well as the physical properties based on these structural models. As the most abundant oxygen-containing groups in GO, epoxy and hydroxyl groups are mainly considered in various structural models of GO. Depending on the coverage and arrangement of these functional groups, the physical properties of GO, such as electronic band gap, electrical conductivity, Young's modulus, and optical transmittance, are tunable in wide ranges.

Chapter 4 summarizes the application of GO in electronics and optics. By appropriately controlling the deposition and reduction parameters, GO films can be made insulating, semiconducting, or semimetallic, while maintaining optical transparency. The tunable electronic and optical properties of GO films lead to some exciting applications in the fields of transparent conductors, field-effect devices, flexible electronic materials, surface-enhanced Raman scattering, optical sensing/detecting, etc.

Chapter 5 describes the great potential of GO and GO-based composites in energy storage and conversion. The tunable electronic properties render GO and GO-based composites excellent catalysts for light-driven hydrogen production from water splitting. Also, their high surface area or porous configurations together with the functional groups contribute to the dehydrogenation of hydrides as well as the physisorption of molecular hydrogen. Moreover, moderate functional groups on GO can immobilize various active species, render various hybrid architectures and simultaneously retain the electrical conductivity for electrode materials in both lithium batteries and supercapacitors.

Chapter 6 depicts the utility of GO and GO composites in air pollutant removal and wastewater treatment. The oxygenated functional groups on the basal plane and edges make GO capable to covalently and noncovalently interact with different molecules. Moreover, the high surface area and functional groups endow GO-based materials great capability to capture the heavy metal ions and organic species. Owing to surface chemistry and architectures, GO-based materials can also function as excellent catalysts or further hybridize with the effective catalysts for converting the harmful gases and organic species in wastewater.

Chapter 7 addresses the application of GO in biotechnology. Benefiting from functional groups such as hydroxyl, epoxy, and carboxyl, GO is allowed to noncovalently interact with biomolecules through electrostatic interaction, π–π stacking, and hydrogen bonding. Moreover, the excellent optical and electromechanical properties of GO extend its applications in biotechnology, especially as a biosensor, which can be used to detect enzyme, DNA, and other biomolecules with high sensitivity and selectivity.

Chapter 8 briefly summarizes the structures, physical properties, and applications of GO and gives some outlooks of this field from the authors.

Acknowledgments

This book was financially supported by the National Natural Science Foundation of China (11134005, 11174045, and 51101145) and the Fundamental Research Funds for the Central Universities of China (5007-852006). The authors thank Prof. S.B. Zhang and Dr. Y.Y. Sun (Rensselaer Polytechnic Institute, USA), Prof. Z.F. Chen and Dr. F.Y. Li (University of Puerto Rico, USA), Dr. L. Wang (Suzhou University, China), Prof. L.X. Sun (Guilin University of Electronic Technology, China), Prof. R. Ahuja (Uppsala University, Sweden), Prof. X.F. Gao (Institute of High Energy Physics, Chinese Academy of Sciences, China), Dr. X. Jiang, and Dr. Y. Su (Dalian University of Technology, China) for collaboration on the graphene oxide and related projects. We also deeply thank Dr. Jian Li in Springer for inviting us to write this book and for many helpful advices during the preparation of the book.

Contents

About the Authors

Dr. Jijun Zhao was born in Jiangsu, China, in 1973. He received his Ph.D. in Condensed Matter Physics from Nanjing University in 1996 and became a professor in Dalian University of Technology in 2006. He is now director of the key laboratory of materials modification by laser, ion, and electron beams (Ministry of Education). His major research field is computational materials science with special interest in clusters, nanostructures, and new energy materials. He has contributed over 300 refereed journal papers and his current H-index is 43. As members in Prof. Zhao's group, **Dr. Lizhao Liu** is a lecturer and **Ms. Fen Li** is a doctoral student in Dalian University of Technology. Lizhao Liu's major interest is carbon nanostructures, and Fen Li is mainly working on hydrogen storage materials.

Chapter 1
Fabrication and Reduction

Abstract GO is a monolayer of graphite oxide. Graphite oxide was successfully fabricated in laboratory for over one and half centuries ago. Generally, there are three main approaches to synthesize graphite oxide. By exfoliating graphite oxide into monolayered structures, GO sheets can be obtained. The formation mechanism of GO from oxidizing graphene will also be discussed. On the other hand, GO is an important raw material to mass-produce graphene via removing the oxygen-containing groups, i.e. reduction of GO. Among different reduction methods, the chemical reduction can occur at low or moderate temperatures with the help of reducing reagents, which is one of the most commonly used approaches to transform GO back into graphene.

1.1 Experimental Fabrication

Similar to graphene (monolayer of graphite), GO can be considered as a monolayer of graphite oxide. As for the graphite oxide, its fabrication can be traced back to more than one and half centuries ago [1, 2]. Early in 1859, Brodie reported synthesis of graphite oxide during investigating the structure of graphite [3]. By treating graphite with the mixture of $KClO_3$ and HNO_3 at 60 °C for three to four days, he found that the product had an increased overall mass. Further analysis showed the product was composed of carbon, hydrogen, and oxygen, with a measured molecular formula of $C_{2.19}H_{0.80}O_{1.00}$. After heating to a temperature of 220 °C, the chemical composition of this material changed to $C_{5.51}H_{0.48}O_{1.00}$, coupled with a loss of carbonic acid and carbonic oxide. Since the material was dispersible in pure or basic water, but not in acidic media, Brodie termed this material "graphic acid".

Later in 1898, Staudenmaier [4] improved Brodie's approach by adding the chlorate (such as $KClO_3$ or $NaClO_3$) in multiple aliquots during the reaction, as well as adding the concentrated H_2SO_4 to increase the acidity of the mixture. The graphite was then treated with the chlorate, H_2SO_4, and HNO_3. Such slight change in the procedure resulted in an overall extent of oxidation similar to

© The Author(s) 2015
J. Zhao et al., *Graphene Oxide: Physics and Applications*,
SpringerBriefs in Physics, DOI 10.1007/978-3-662-44829-8_1

Brodie's multiple oxidation approach (C:O ~ 2:1), but performed more practically in a single reaction vessel.

In 1958, Hummers and Offeman [5] developed an alternative oxidation method by reacting graphite with a mixture of $KMnO_4$ and concentrated H_2SO_4, which also achieved similar levels of oxidation. Graphite oxide was prepared by mixing ultrapure graphite powder and $NaNO_3$ in H_2SO_4. Then $KMnO_4$ was added to catalyze the reaction and finally a brownish grey gel was obtained. After water diluting and hydrogen peroxide treatment, a yellow-brown graphite oxide residue was obtained, which can be used to fabricate graphite oxide suspension with aid of sonication. Though permanganate is a commonly used oxidant, the real active species is Mn_2O_7 [1], which is able to selectively oxidize unsaturated aliphatic double bonds over aromatic double bonds.

Generally speaking, Brodie, Staudenmaier, and Hummers are the three major methods to produce graphite oxide from graphite. However, both the Brodie and Staudenmaier methods generate ClO_2 gas, which must be handled with caution due to its high toxicity and tendency to decompose in air to produce explosions. Fortunately, this drawback can be eliminated by the Hummers method, which has relatively shorter reaction time and is absent of hazardous ClO_2. Nowadays, the Hummers method has been widely used, with still one drawback, that is, the potential contamination by excess permanganate ions. Such contamination can be removed by H_2O_2 treatment, followed by washing and thorough dialysis [6]. On the other hand, it has been demonstrated that the products of graphite oxide synthesis reactions show strong variance, depending on not only the particular oxidants used, but also the graphite source and reaction conditions.

After fabrication of graphite oxide, GO can be obtained by exfoliating graphite oxide into monolayer sheets through a variety of thermal and mechanical methods [1]. Early in 1962, Boehm et al. pioneered the thermal exfoliation of graphite oxide, where thin layered carbon films were prepared by thermal deflagration of graphite oxide [7]. Afterwards, thermal exfoliation gradually becomes a popular method to peel graphite oxide to achieve graphene [8, 9]. According to McAllister et al.'s report [10], the mechanism of thermal exfoliation of graphite oxide can be described in Fig. 1.1. During heating, the oxygen-containing functional groups attached on carbon plane decompose into gases such as H_2O, CO_2, and CO, which will diffuse along the lateral direction; the exfoliation occurs only if the decomposition rate of functional groups surpasses the diffusion rate of evolved gases. In the case, the interlayer pressure existing among adjacent layers is large enough to overcome their van der Waals interactions and pushes the layers separated from each other. Generally, a minimum temperature of 550 °C was suggested to be necessary for the successful exfoliation at atmospheric pressure [10]. Moreover, the method of thermal exfoliation has been widely used to obtain graphene, as illustrated in Fig. 1.1, where graphite oxide was thermal exfoliated into graphene [11].

Another conventional way of converting from graphite oxide to GO is to mechanically exfoliate graphite oxide. For example, by sonicating graphite oxide in water or polar organic media, one can completely exfoliate the graphite oxide into GO [12–15]. Besides, through mechanically stirring graphite oxide in water,

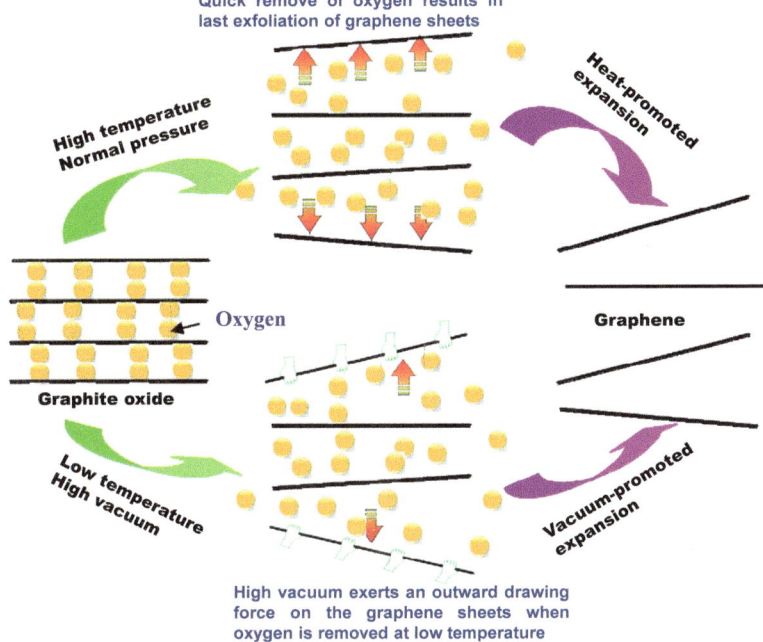

Fig. 1.1 Schematic showing thermal exfoliation of graphite oxide. Reprinted with permission from Ref. [11]. Copyright (2009) American Chemical Society

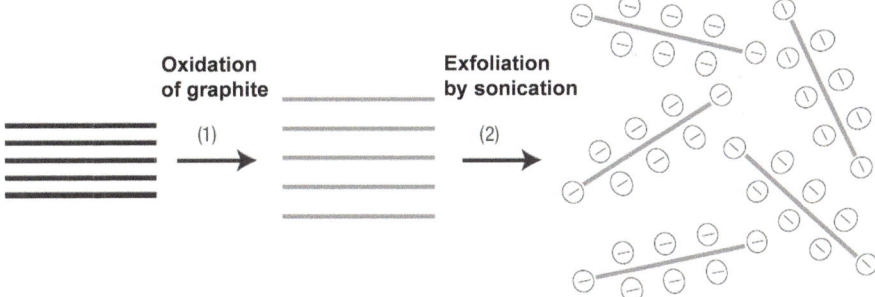

Fig. 1.2 Schematic showing the exfoliation of graphite oxide in water by sonication. Reprinted with permission from Ref. [18]. Copyright (2008) Nature Publishing Group

graphite oxide can be also well exfoliated [16, 17]. Especially, sonicating and mechanical stirring can be combined together to exfoliate graphite oxide with a better efficiency than using any individual method. Generally speaking, sonicating graphite oxide in water or polar organic media is much faster compared with the mechanical stirring, but it has a great disadvantage in causing substantial damage to the GO platelets [15]. Figure 1.2 displays schematically the mechanical exfoliation of graphite oxide by sonication.

1.2 Oxidation Mechanism

Motivated by the fabrication of GO, experiments have also been carried out to illustrate the oxidation mechanism of graphite. As mentioned above, during fabricating GO, the graphite was reacted with acids, which are strong oxidizing agents. Since the edges of graphite are open and the distance between graphitic layers is about 0.334 nm, an easily conceived oxidation mechanism of acid penetration into graphite is an "interlayer-acid-penetration" (acid-intercalation) mechanism [19, 20]. During intercalation with the acid oxidant, potential of graphite increases with oxidation and there is an upper limit of saturated potential, depending strongly on the oxidizer and the concentration of acid used [20]. Recently, combining several spectroscopic techniques, Shin et al. [21] were able to elucidate the oxidation mechanism of highly ordered pyrolytic graphite (HOPG) in a nitric acid/sulfuric acid mixture, as shown in Fig. 1.3. They found a new corrosion process of "direct-acid-penetration" from the outer to inner graphitic layers with nitration and sulfonation, suggesting that the acid oxidant not only introduces oxygen-containing groups together with nitrogen- and sulfur-containing groups, but also creates structural defects in the graphitic lattice.

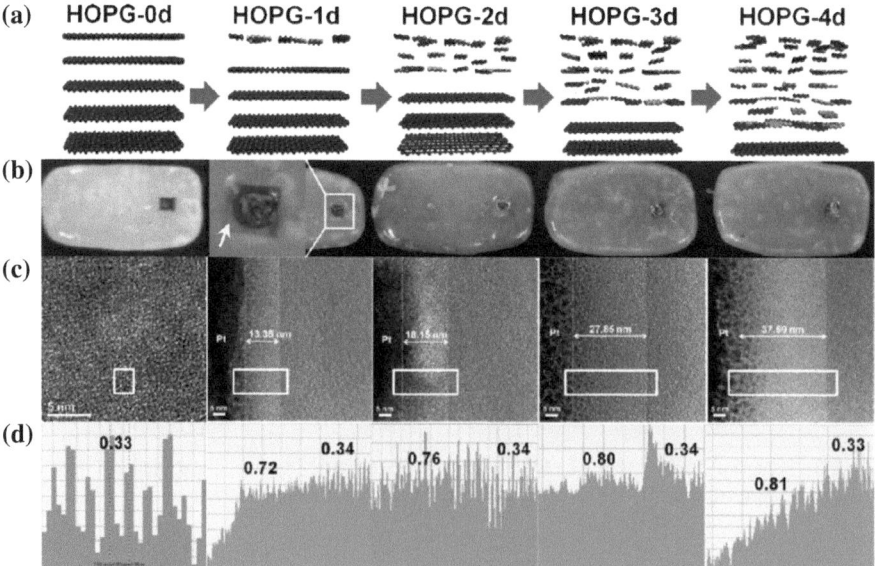

Fig. 1.3 **a** Schematic showing destructive oxidation for HOPG in a nitric acid/sulfuric acid mixture at 100 °C with respect to treating time from 0 to 4 days; **b** photographs of sample specimens with an *inset* showing the stability of molten high-density polyethylene in a nitric acid/sulfuric acid at 100 °C for 1 day; **c** TEM images of the cross-sectioned samples with the increased depth illustrating the oxidation and structural etching progress; **d** layer distance obtained from *rectangles* for TEM images in (**c**). Reprinted with permission from Ref. [21]. Copyright (2012) Elsevier Ltd.

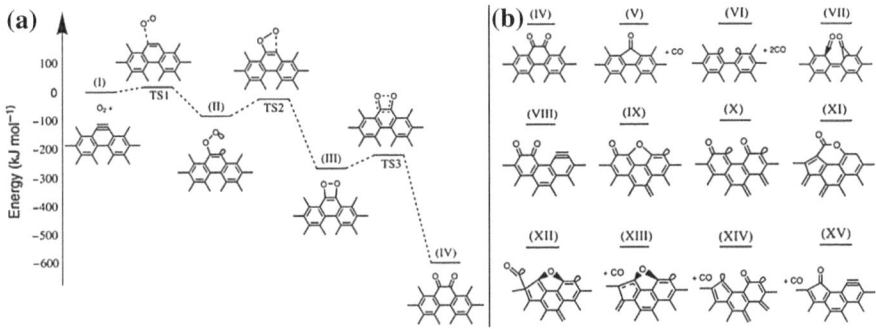

Fig. 1.4 **a** Potential energy surface for O_2 chemisorption; **b** different metastable structures from IV to XV. Reprinted with permission from Ref. [24]. Copyright (2004) The Combustion Institute

Parallel to the experimental investigations, theoretical calculations with density functional theory (DFT) have also been performed to elucidate the oxidation mechanism of graphene. Since the oxidation mechanism in realistic process of GO synthesis is very complicated, theoretical studies mainly focused on simple processes, such as the reaction of O_2 with graphene. In general, there are several oxidation passways for O_2 reacting with the graphene [22, 23], which are addressed in the following:

(1) **Reaction of O_2 with the graphene armchair edges**. Sendt and Haynes [24] studied oxidation on the dehydrogenated armchair edge of graphene and found that chemisorption of O_2 on the armchair edges is thermodynamically favorable to form two quinone structures with a small energy barrier (18 kJ/mol) and a large exothermic heat of reaction (−578 kJ/mol), as shown in Fig. 1.4a. Afterwards, a series of reactions can further contribute to carbon gasification, such as desorption, rearrangement, and migration. At low temperatures, due to low internal energy, dissipation and low-barrier reactions readily occur, allowing oxide structures such as (IV), (IX), and (X) to be stabilized. At intermediate temperatures, products are likely to be (V), (VII), and any products from reactions of (X) with barriers <300 kJ/mol. At high temperatures, the carbon possesses sufficient internal energy to allow reactions with high barriers, and products such as (VI), (XIV), and (XV) appear. These structures are shown in Fig. 1.4b.

(2) **Reaction of O_2 with the graphene zigzag edges**. Zhu et al. [25] simulated the reaction of O_2 with graphene zigzag edges. Owning to the large exothermicity for O_2 chemisorption on the bare edges, nearly all edge sites can chemisorb oxygen atoms. As a consequence, a large number of semi-quinone and *o*-quinone oxygen can be formed, indicating a significant increase of the number of active sites. Moreover, the weaker *o*-quinone C–C bonds also drive the reaction forward at (ca. 30 %) lower activation energy. Epoxy oxygen forms under relatively high O_2 pressure; it can only increase the number of active sites without further reducing the activation energy.

(3) **Reaction of O_2 with the graphene vacancies**. For reaction of O_2 with the graphene armchair or zigzag edges, CO is the main product during gasification process. However, it was CO_2 rather than CO that was experimentally observed for complete combustion under atmospheric oxygen pressure. To explain this phenomenon, Carlsson et al. [26] examined the oxidation of graphene vacancies and proposed a two-step oxidation mechanism at low temperature. First, vacancies are initially saturated by stable O-containing groups such as ether (C–O–C) and carbonyl (C=O). The second step is an etching process activated by additional O_2 adsorption at the ether groups. Larger oxygen-containing groups like lactone (C–O–C=O) and anhydride (O=C–O–C=O) are formed, which may desorb as CO_2 just above room temperature.

(4) **Reaction of O_2 with the graphene basal plane**. Lahaye et al. [27] revealed that the oxygen atoms and the adjacent carbon atoms form epoxide groups on the carbon lattice. During the oxidation process, the hydroxyl groups are formed on the opposite side of the basal plane to relax the tension induced by the epoxide groups. In addition, line oxidation defects have been observed in the experimentally fabricated highly oriented pyrolytic graphite (HOPG). To realize this kind of defects, Li et al. [28] further illustrated the graphene oxidative breakup process, which provides a useful insight into this puzzling GO structure. At first, a single O atom is added to coronene, forming an epoxy group with an exothermicity of ~2.4 eV, as shown in Fig. 1.5a. At the same time, the original epoxy C–C bond (1.42 Å) is stretched to 1.58 Å. Then, addition of the second O atom is preferable at the opposite carbon site of the epoxy group (Fig. 1.5b). This unzipping configuration is thermodynamically more stable than any other configurations by ~1.2 eV. Further adding O atoms will gradually unzip the graphene, as shown in Fig. 1.5c–d.

Fig. 1.5 Coronene molecules as a prototype of graphene fragments attached with **a** one, **b** two, **c** three and **d** four epoxy groups, respectively. Reprinted with permission from Ref. [28]. Copyright (2006) American Physical Society

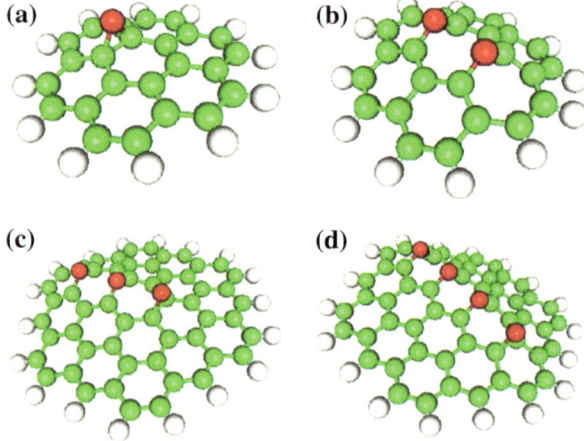

1.3 Chemical Reduction

Usually, GO is thermally unstable and can be easily (partly) reduced by removing the oxygen-containing groups. The reduced GO (RGO) sheets are usually considered as one kind of chemically derived graphene. Also, reduction of GO is an important approach to produce graphene. Compared with GO, RGO has lower C:O ratio due to losing of oxygen-containing groups. Depending on the fabrication methods, typically produced GO has chemical compositions ranging from $C_8O_2H_3$ to $C_8O_4H_5$, corresponding to a C:O ratio of 4:1–2:1 [8]. After reduction, the C:O ratio can be increased to ~12:1 in most cases [29, 30], and even as large as 246:1 [31]. Generally, there are three kinds of reduction approaches of GO, including thermal reduction, chemical reduction, and multi-step reduction. The thermal reduction can be achieved by thermal irradiation [16, 29, 30, 32, 33], microwave irradiation [34, 35], photon irradiation [36, 37], and other unconventional heating resources [38, 39]. The chemical reduction is usually aided by chemical agents or other chemical processes to remove the oxygen-containing groups. Both thermal reduction and chemical reduction are mostly realized in one step. However, the multi-step reduction can further improve the reduction efficiency for some special purposes. A typical multi-step reduction is the three-step process proposed by Gao et al. [31], which includes steps of de-oxygenation with $NaBH_4$, de-hydration with concentrated sulfuric acid and thermal annealing. Compared with the other reduction methods, the chemical reduction can occur at low or moderate temperatures with the aid of reducing reagents. In the following, we will illustrate several chemical reduction ways in details.

1.3.1 Ways of Chemical Reduction

A common way of chemical reduction is the chemical reagent reduction, where GO is reduced by chemical agents [2]. So far, various chemical agents have been used, including hydrazine, alcohol, sodium borohydride, hydriodic acid with acetic acid, sodium/potassium hydroxide, iron/aluminium powder, ammonia, hexylamine, sulfur-containing compounds ($NaHSO_3$, Na_2SO_3, $Na_2S_2O_4$, $Na_2S_2O_3$, $Na_2S \cdot 9H_2O$, $SOCl_2$, and SO_2), hydroxylamine hydrochloride, urea, lysozyme, vitamin C, N-methyl-2-pyrrolidinone, poly(norepinephrine), BSA, TiO_2 nanoparticles, manganese oxide, and bacteria respiration [40]. Among them, the first and most commonly used was hydrazine (N_2H_4), which is able to produce highly reduced graphene oxide under low temperature. With hydrazine hydrate, Stankovich et al. [13, 41] reported the first reduction of a colloidal suspension of exfoliated GO sheets in water, resulting in the aggregation and subsequent formation of a high-surface-area carbon material that consists of thin graphene-based sheets. In this method, reduction of GO was achieved by adding hydrazine hydrate into the GO suspension and heating in an oil bath at 100 °C under a water-cooled condenser for 24 h. Nowadays, the highest conductivity of RGO films produced solely by hydrazine reduction is 99.6 S/cm, which has a C:O ratio of around 12.5 [8]. However, hydrazine is toxic, preventing its wide usage in mass production of RGO. To avoid this drawback, many other "non-toxic" reducing agents, such as alcohol, sodium/potassium hydroxide, sodium borohydride, hydriodic acid, and acetic acid, are employed.

Another chemical reduction method is photocatalyst reduction with the assistance of a photocatalyst like TiO_2. With the assistance of TiO_2 particles under ultraviolet (UV) irradiation, GO in a colloid state changes in color from light brown to dark brown to black, which can be seen as a sign of GO reduction, as shown in Fig. 1.6 [42]. Such color change was explained by partial restoration of the conjugated network in the carbon plane. In addition, this photocatalyst reduction can be ascribed to two chemical processes, which are described in the formula in Fig. 1.6. Upon UV-irradiation, charge separation occurs on the surface of TiO_2 particles. In the presence of ethanol, the holes are scavenged to produce ethoxy radicals, which in turn leave the electrons to accumulate within the TiO_2 particles. The accumulated electrons interact with GO sheets to reduce the functional groups.

Furthermore, oxygen-containing groups of GO can be removed by electrochemical reduction [43–46]. Electrochemical reduction of GO are carried out in a normal electrochemical cell using an aqueous buffer solution at room temperature. The reduction usually needs no special chemical agent, and is mainly caused by the electron exchange between GO and electrodes. Therefore, electrochemical reduction is advantageous to avoid the usage of dangerous reductants (e.g. hydrazine) and to eliminate byproducts. In 2009, Ramesha and Sampath [43] studied the electrochemical reduction of GO. After depositing a thin film of GO on a substrate (glass, plastic, ITO, etc.), an inert electrode was placed oppositely the film in an electrochemical cell and the reducing process occurred during charging of the cell.

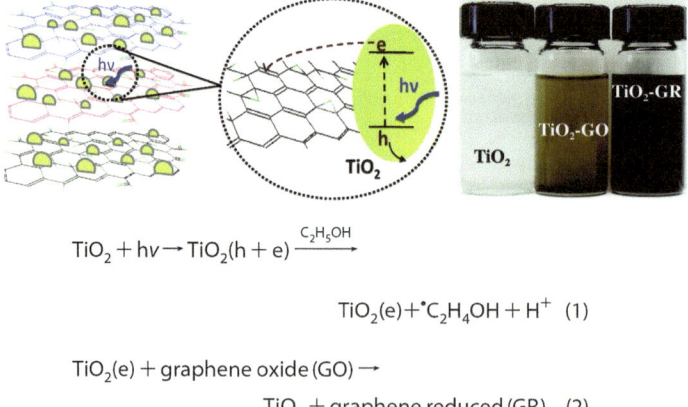

$$TiO_2 + h\nu \rightarrow TiO_2(h+e) \xrightarrow{C_2H_5OH}$$

$$TiO_2(e) + {}^{\bullet}C_2H_4OH + H^+ \quad (1)$$

$$TiO_2(e) + graphene\ oxide\ (GO) \rightarrow$$

$$TiO_2 + graphene\ reduced\ (GR) \quad (2)$$

Fig. 1.6 Change in color of GO showing its photocatalyst reduction with the aid of TiO_2, as well as the two main chemical processes during photocatalyst reduction. Reprinted with permission from Ref. [42]. Copyright (2008) American Chemical Society

They found that the reduction of GO began at -0.6 V and reached a maximum at -0.87 V. In an experiment by Zhou et al. [44], RGO sample reaches a C:O ratio of 23.9 and a conductivity of 85 S/cm. Moreover, the reduction can be controlled by the pH value of the buffer solution. A low pH value is favorable to reduce GO, indicating that H^+ ions may participate in the reaction.

1.3.2 Theoretical Illustration of GO Reduction

Employing DFT calculations, Kim et al. [47] examined the epoxide reduction with hydrazine on a single-layer graphene and pointed out that the reduction follows a direct Eley–Rideal mechanism rather than a Langmuir–Hinshelwood mechanism. The reduction reaction is mainly governed by epoxide ring opening, which is initiated by hydrogen transfer from hydrazine or its derivatives. The formation of derivatives such as $NHNH_2$ during reduction can facilitate the de-epoxidation by lowering the barrier height of the ring-opening reaction.

Gao et al. [48] further elucidated the effect of hydrazine treatment on different functional groups and proposed several reduction routes for de-epoxidation by hydrazine, where all routes start from ring-opening of epoxy groups and form hydroxyl groups on the original sites, as illustrated in Fig. 1.7. It was shown that the hydrazine reduction can only reduce epoxy groups, while no reaction path was found for the reduction of hydroxyl, carbonyl and carboxyl groups of GO. Generally, the mechanism of hydrazine reduction can be addressed as follows: the reduction is postulated to proceed via a direct nucleophilic attack of hydrazine on an epoxide group to result in a hydrazine alcohol moiety, which releases a water molecule towards the formation of an aminoaziridine and finally undergo a thermal elimination of diimide to form a double bond [2].

In addition to the nitrogen-containing reducing agents (e.g. hydrazine), sulphur-containing reducing agents in reduction of GO have also been widely used. Employing the ab initio calculations, Su et al. [49] studied the reaction mechanism of chemical reduction of GO by the sulfur-containing agents, i.e. HSO_3^- and H_2SO_3. Three reaction processes were discussed: de-hydroxylation of GO with one or two hydroxyl groups, de-epoxidation of GO with one or two epoxy groups and de-carboxylation, and de-carbonylation of GO with carboxyl and carbonyl groups. It was found that hydroxyl and epoxide groups could be easily reduced owing to lower energy barriers, whereas de-carboxylation and de-carbonylation reactions are not kinetically and thermodynamically easy due to higher energy barriers. As illustrated in Fig. 1.8a, for reduction of the hydroxyl group, the HSO_3^- ion can approach the hydroxyl group through electrostatic attraction between the HSO_3^- anion and the partially positively charged H in the hydroxyl group, finally releasing an exothermicity of 20.3 kcal/mol. However, without the aid of HSO_3^- anion, it will be more difficult to remove the hydroxyl group directly (Fig. 1.8b). Similarly, for reduction of the epoxide group, the H atom of HSO_3^- anion will attack the epoxy group and open the epoxide ring. As a result,

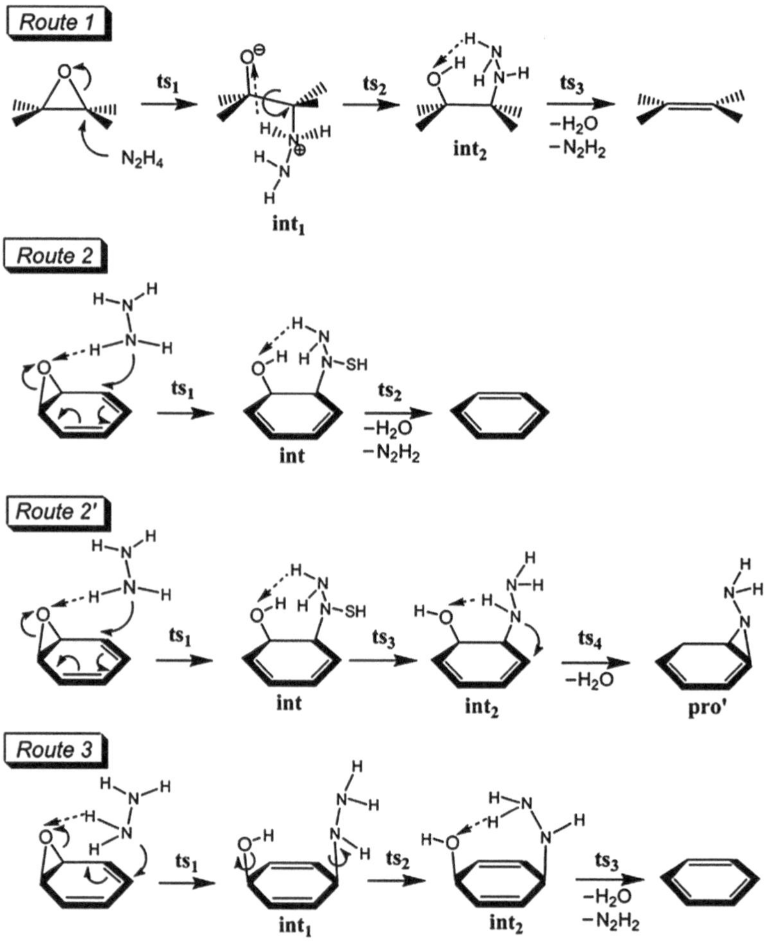

Fig. 1.7 The mechanisms for the hydrazine de-epoxidation of GO. Reprinted with permission from Ref. [48]. Copyright (2010) American Chemical Society

the H atom is transferred from HSO_3^- to the epoxide group, forming intermediates of hydroxyl group and SO_3, as shown in Fig. 1.8c. In the following, the H_2SO_3 can further react with the epoxide group and form a H_2SO_4, as shown in Fig. 1.8d. In this case, the S atom will react with the epoxide group and open the epoxide ring. The above reduction mechanism of epoxide group is experimentally supported by Chen et al. [50], who also suggested that the during reduction of the epoxide group, the S atom of the sulphur-containing ion will attack the epoxide group and break the C–O bond, finally obtaining the product of H_2SO_4.

Another kind of reducing agents are metal-containing agents. Generally, there are two modes of mechanisms of this kind of reduction. One is fast electron transfer between metal and GO. The other is evolution of nascent hydrogen as the active reducing agent [2]. For example, Xie et al. [51] investigated the reduction

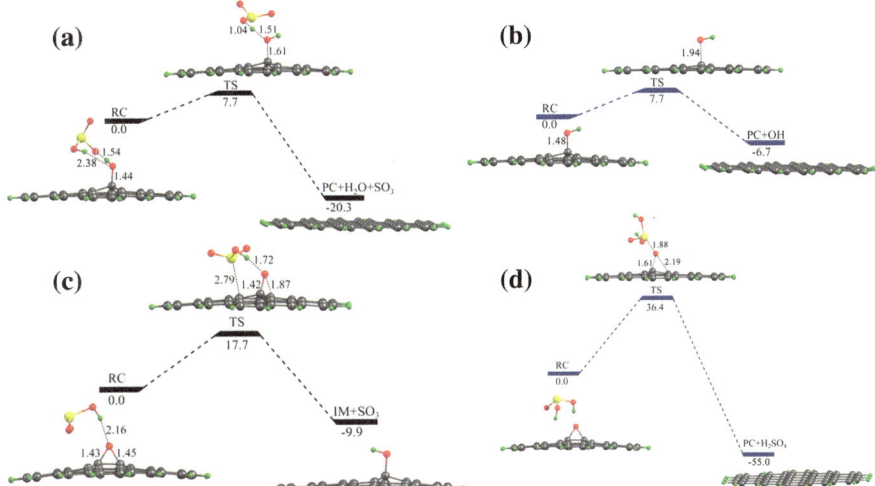

Fig. 1.8 Relative Gibbs free energy profile of HSO₃⁻ de-hydroxylation (**a**) and direct removal of the hydroxyl group (**b**) on GO with only one hydroxyl group; Relative Gibbs free energy profile of HSO₃⁻ (**c**) and H₂SO₃ (**d**) de-epoxidation on GO with only one epoxide group. Reprinted with permission from Ref. [49]. Copyright (2013) Elsevier Ltd.

of GO by alkali-earth metals. They found that alkali-earth metals are highly effective in reducing GO not only because it can accelerate oxygen desorption but also because it can eliminate carbonyl defects to recover perfect graphene lattice. The physical mechanism for alkali earth metal assisted GO reduction is identified as the relaying of three charge transfer steps: (a) from Ca to pristine carbon empty states, (b) from pristine carbon to oxidized carbon, and (c) from oxidized carbon to oxygen at the reaction site, respectively.

References

1. Dreyer, D.R., Park, S., Bielawski, C.W., Ruoff, R.S.: Chem. Soc. Rev. **39**, 228–240 (2010)
2. Chua, C.K., Pumera, M.: Chem. Soc. Rev. **43**, 291–312 (2014)
3. Brodie, B.C.: Philos. Trans. R. Soc. Lond. **149**, 249–259 (1859)
4. Staudenmaier, L.: Ber. Dtsch. Chem. Ges. **31**, 1481–1487 (1898)
5. Hummers, W.S., Offeman, R.E.: J. Am. Chem. Soc. **80**, 1339 (1958)
6. Johnson, J.A., Benmore, C.J., Stankovich, S., Ruoff, R.S.: Carbon **47**, 2239–2243 (2009)
7. Boehm, H.-P., Clauss, A., Fischer, G.O., Hofmann, U.Z.: Anorg. Allg. Chem. **316**, 119–127 (1962)
8. Pei, S., Cheng, H.M.: Carbon **50**, 3210–3228 (2012)
9. Botas, C., Álvarez, P., Blanco, C., Santamaría, R., Granda, M., Gutiérrez, M.D., Rodríguez-Reinoso, F., Menéndez, R.: Carbon **52**, 476–485 (2013)
10. McAllister, M.J., Li, J.-L., Adamson, D.H., Schniepp, H.C., Abdala, A.A., Liu, J., Herrera-Alonso, M., Milius, D.L., Car, R., Prud'homme, R.K.: Chem. Mater. **19**, 4396–4404 (2007)
11. Lv, W., Tang, D.M., He, Y.B., You, C.H., Shi, Z.Q., Chen, X.-C., Chen, C.-M., Hou, P.-X., Liu, C., Yang, Q.H.: ACS Nano **3**, 3730–3736 (2009)

12. Stankovich, S., Piner, R.D., Nguyen, S.T., Ruoff, R.S.: Carbon **44**, 3342–3347 (2006)
13. Stankovich, S., Piner, R.D., Chen, X., Wu, N., Nguyen, S.T., Ruoff, R.S.: J. Mater. Chem. **16**, 155–158 (2006)
14. Stankovich, S., Dikin, D.A., Dommett, G.H.B., Kohlhaas, K.M., Zimney, E.J., Stach, E.A., Piner, R.D., Nguyen, S.T., Ruoff, R.S.: Nature **442**, 282–286 (2006)
15. Paredes, J.I., Villar-Rodil, S., Martínez-Alonso, A., Tascón, J.M.D.: Langmuir **24**, 10560–10564 (2008)
16. Becerril, H.A., Mao, J., Liu, Z., Stoltenberg, R.M., Bao, Z., Chen, Y.: ACS Nano **2**, 463–470 (2008)
17. Zhu, Y., Stoller, M.D., Cai, W., Velamakanni, A., Piner, R.D., Chen, D., Ruoff, R.S.: ACS Nano **4**, 1227–1233 (2010)
18. Li, D., Müller, M.B., Gilje, S., Kaner, R.B., Wallace, G.G.: Nat. Nanotechnol. **3**, 101–105 (2008)
19. Forsman, W.C., Vogel, F.L., Carl, D.E., Hoffman, J.: Carbon **16**, 269–271 (1978)
20. Inagaki, M., Iwashita, N., Kouno, E.: Carbon **28**, 49–55 (1990)
21. Shin, Y.R., Jung, S.M., Jeon, I.Y., Baek, J.B.: Carbon **52**, 493–498 (2013)
22. Gao, X., Jiang, D.E., Zhao, Y., Nagase, S., Zhang, S., Chen, Z.: J. Comput. Theor. Nanosci. **8**, 2406–2422 (2011)
23. Lu, N., Li, Z.: Graphene oxide: theoretical perspectives. In: Zeng, J., Zhang, R.Q., Treutlein, H.R. (eds.) Quantum Simulations of Materials and Biological Systems, pp. 69–84. Springer, Dordrecht (2012)
24. Sendt, K., Haynes, B.S.: Proc. Combust. Inst. **30**, 2141–2149 (2005)
25. Zhu, Z.H., Finnerty, J., Lu, G.Q., Yang, R.T.: Energy Fuels **16**, 1359–1368 (2002)
26. Carlsson, J.M., Hanke, F., Linic, S., Scheffler, M.: Phys. Rev. Lett. **102**, 166104 (2009)
27. Lahaye, R.J.W.E., Jeong, H.K., Park, C.Y., Lee, Y.H.: Phys. Rev. B **79**, 125435 (2009)
28. Li, J.-L., Kudin, K.N., McAllister, M.J., Prud'homme, R.K., Aksay, I.A., Car, R.: Phys. Rev. Lett. **96**, 176101 (2006)
29. Schniepp, H.C., Li, J.L., McAllister, M.J., Sai, H., Herrera-Alonso, M., Adamson, D.H., Prud'homme, R.K., Car, R., Saville, D.A., Aksay, I.A.: J. Phys. Chem. B **110**, 8535–8539 (2006)
30. Mattevi, C., Eda, G., Agnoli, S., Miller, S., Mkhoyan, K.A., Celik, O., Mastrogiovanni, D., Granozzi, G., Garfunkel, E., Chhowalla, M.: Adv. Funct. Mater. **19**, 2577–2583 (2009)
31. Gao, W., Alemany, L.B., Ci, L.J., Ajayan, P.M.: Nat. Chem. **1**, 403–408 (2009)
32. Yang, D., Velamakanni, A., Bozoklu, G., Park, S., Stoller, M., Piner, R.D., Stankovich, S., Jung, I., Field, D.A., Ventrice Jr, C.A., Ruoff, R.S.: Carbon **47**, 145–152 (2009)
33. Wang, X., Zhi, L., Mullen, K.: Nano Lett. **8**, 323–327 (2007)
34. Zhu, Y., Murali, S., Stoller, M.D., Velamakanni, A., Piner, R.D., Ruoff, R.S.: Carbon **48**, 2118–2122 (2010)
35. Hassan, H.M.A., Abdelsayed, V., Abd El Rahman, S.K., AbouZeid, K.M., Terner, J., El-Shall, M.S., Al-Resayes, S.I., El-Azhary, A.A.: J. Mater. Chem. **19**, 3832–3837 (2009)
36. Cote, L.J., Cruz-Silva, R., Huang, J.: J. Am. Chem. Soc. **131**, 11027–11032 (2009)
37. Zhang, Y., Guo, L., Wei, S., He, Y., Xia, H., Chen, Q., Sun, H.-B., Xiao, F.-S.: Nano Today **5**, 15–20 (2010)
38. Baraket, M., Walton, S.G., Wei, Z., Lock, E.H., Robinson, J.T., Sheehan, P.: Carbon **48**, 3382–3390 (2010)
39. Guo, Y., Wu, B., Liu, H., Ma, Y., Yang, Y., Zheng, J., Yu, G., Liu, Y.: Adv. Mater. **23**, 4626–4630 (2011)
40. Mao, S., Pu, H., Chen, J.: RSC Adv. **2**, 2643–2662 (2012)
41. Stankovich, S., Dikin, D.A., Piner, R.D., Kohlhaas, K.A., Kleinhammes, A., Jia, Y., Wu, Y., Nguyen, S.T., Ruoff, R.S.: Carbon **45**, 1558–1565 (2007)
42. Williams, G., Seger, B., Kamat, P.V.: ACS Nano **2**, 1487–1491 (2008)
43. Ramesha, G.K., Sampath, S.: J. Phys. Chem. C **113**, 7985–7989 (2009)

44. Zhou, M., Wang, Y., Zhai, Y., Zhai, J., Ren, W., Wang, F., Dong, S. Chem.-A Eur. J. **15**, 6116–6120 (2009)
45. Wang, Z., Zhou, X., Zhang, J., Boey, F., Zhang, H.: J. Phys. Chem. C **113**, 14071–14075 (2009)
46. An, S.J., Zhu, Y., Lee, S.H., Stoller, M.D., Emilsson, T., Park, S., Velamakanni, A., An, J., Ruoff, R.S.: J. Phys. Chem. Lett. **1**, 1259–1263 (2010)
47. Kim, M.C., Hwang, G.S., Ruoff, R.S.: J. Chem. Phys. **131**, 064704 (2009)
48. Gao, X., Jang, J., Nagase, S.: J. Phys. Chem. C **114**, 832–842 (2010)
49. Su, Y., Gao, X., Zhao, J.: Carbon **67**, 146–155 (2014)
50. Chen, W., Yan, L., Bangal, P.R.: J. Phys. Chem. C **114**, 19885–19890 (2010)
51. Xie, S.Y., Li, X.B., Sun, Y.Y., Zhang, Y.L., Han, D., Tian, W.Q., Wang, W.Q., Zheng, Y.S., Zhang, S.B., Sun, H.-B.: Carbon **52**, 122–127 (2013)

Chapter 2
Structural Characterization

Abstract The experimentally fabricated GO powders are usually water dispersible, insulating, and light brown in color. After the experimental synthesis, one important issue is to determine the structure of GO. So far, the structure of GO is still ambiguous due to its nonstoichiometry since the types of oxygen-containing groups and their arrangements across the carbon network vary substantially under different synthesis conditions. Generally, to determine the structure of GO, some primary questions have to be illuminated: (1) Which functional groups are present? (2) What are the amount and relative fraction of the functional groups? (3) How do these functional groups distribute spatially over the graphene plane? (4) How do the amount and distribution of these groups evolve during reduction? Spectroscopic techniques, such as solid-state nuclear magnetic resonance (NMR), X-ray photoelectron spectroscopy (XPS), X-ray absorption near-edge spectroscopy (XANES), Fourier transform infrared spectroscopy (FT-IR) and Raman spectroscopy, can provide essential insights into the types of oxygenated functional groups in GO and their distributions. Besides, microscopic techniques including transmission electron microscopy (TEM), scanning tunneling microscopy (STM), atomic force microscopy (AFM), and scanning transmission electron microscopy (STEM) have also been used to determine the atomic structures of GO.

2.1 Spectroscopic Characterization

2.1.1 Solid-State NMR

The solid-state NMR spectra of different GO samples exhibit similar resonance patterns featuring three peaks at 60, 70 and 130 ppm and their relative intensities do not change significantly upon oxidation. In 1996, Klinowski's group [1] studied the structure of GO by using the ^{13}C and ^{1}H NMR spectra and assigned the 60 ppm peak to hydroxyl (C–OH), the 70 ppm peak to epoxide (C–O–C), and the 130 ppm one to non-aromatic carbon double bonds ($>C=C<$). However, the identification

© The Author(s) 2015
J. Zhao et al., *Graphene Oxide: Physics and Applications*,
SpringerBriefs in Physics, DOI 10.1007/978-3-662-44829-8_2

of the first two peaks at 60 and 70 ppm was uncertain due to lacking of sufficient spectral information. Then, in 1997 and 1998, the same group revisited the structure of GO and critically assigned the peak around 60 ppm to epoxide (C–O–C), the peak around 70 ppm to hydroxyl (C–OH), and the peak around 130 ppm to non-aromatic carbon double bonds (>C=C<) [2, 3]. Afterwards, Szabó et al. [4] also examined the structure of GO by the ^{13}C NMR spectra and obtained two featured signals at 57.6 and 69.2 ppm, which corresponds to the epoxide (C–O–C) and hydroxyl (C–OH), respectively. In addition, they found two other peaks at 92.9 and 166.3 ppm as well, but were not able to clearly assign them to any specific groups.

A more accurate NMR characterization of GO was fulfilled in 2008 by Cai et al. [5] using high-resolution solid-state NMR with magic angle spinning (MAS). It was confirmed that the peak around 60 ppm corresponds to epoxide, the peak around 70 ppm corresponds to hydroxyl, and the peak around 130 ppm corresponds to sp^2 carbon. Until then, the three major chemical shift peaks around 60, 70, and 130 ppm are commonly accepted and assigned to epoxy, hydroxyl and graphitic sp^2 carbon, respectively. As a consequence, epoxy and hydroxyl are determined as the two major functional groups across the basal plane of GO. In addition, in the high-resolution ^{13}C NMR spectra [4–6], the other three minor peaks at about 101, 167, and 191 ppm were also found, which were tentatively assigned to lactol, ester carbonyl and ketone groups, respectively. In 2009, Gao et al. [7] further assigned the peak around 101 ppm to five- or six-membered-ring lactol decorated on the edge of holes in GO flakes.

A recent study by Zhang et al. [8] labeled all the six peaks mentioned above, where peaks at 61, 70, 101, 130, 169 and 193 ppm were assigned to groups of C–O–C, C–OH, O–C–O, graphitic sp^2 C, O=C–O, and C=O, separately. To gain information about the distribution of major functional groups, two- and multi-dimensional NMR spectra conducted by Ruoff's group [5, 9] revealed that epoxide and hydroxyl were close to each other, with some tiny islands of pure epoxies or hydroxyls. Diagonal signals eliminated in the two-dimensional double-quantum/single-quantum spectrum shows that the cross peak of epoxide is at (ω_{SQ}, ω_{DQ}) = (60, 130 ppm) and the cross peak of hydroxyl is at (ω_{SQ}, ω_{DQ}) = (70, 130 ppm), respectively [9]. This confirms clearly that the epoxide and hydroxyl carbons are directly bonded. Besides, the major peaks (60 and 70 ppm) in the NMR spectra were related to the carbon atoms single-bonded to oxygen atoms. Figure 2.1 gives the typical ^{13}C NMR spectra of GO.

2.1.2 XPS

As a complementarity to the NMR spectra, XPS can further unveil the nature of carbon atoms in different chemical environments. Comparing the survey scan spectra of bare graphite and GO [4], it was found that the C and O atoms were presented at the surface of both systems; but the oxygen content of GOs

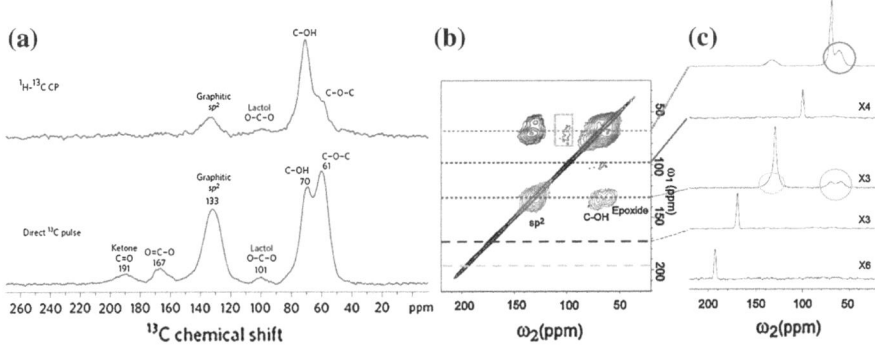

Fig. 2.1 **a** ^{1}H–^{13}C cross polarization (CP) spectrum of GO obtained with 7.6 kHz MAS and a contact time of 1 ms (67,000 scans), and a direct ^{13}C pulse spectrum obtained with 12 kHz MAS and a 90° ^{13}C pulse (10,000 scans). Reprinted with permission from Ref. [7]. Copyright (2009) Nature Publishing Group. **b** 2D ^{13}C/^{13}C chemical-shift correlation solid-state NMR spectra of GO; **c** slices selected from the 2D spectrum at the indicated positions (70, 101, 130, 169, and 193 ppm) in the ω_1 dimension. Reprinted with permission from Ref. [5]. Copyright (2008) American Association for the Advancement of Science

was much higher because of the oxidative treatment, as shown in Fig. 2.2a, b. Further high-resolution XPS spectra demonstrated that in the C 1s signal of pristine GO, there are five different chemically shifted components at 284.5, 285.86, 286.55, 287.5 and 289.2 eV respectively, which can be assigned to sp^2 carbon atoms in aromatic rings (284.5 eV) and C atoms bonded to hydroxyl (C–OH, 285.86 eV), epoxy (C–O–C, 286.55 eV), carbonyl (>C=O, 287.5 eV), and carboxyl groups (COOH, 289.2 eV), respectively [10–13], as shown in Fig. 2.2c. However, the presence of carbonyl (>C=O) groups is still ambiguous. Some reports [3, 14] only considered four feature components of the deconvolution of the C 1s spectra by ignoring the presence of >C=O groups, i.e. sp^2 carbons, C–OH, C–O–C, and COOH. Further information provided by the O 1s spectra can complement the information from the C 1s spectra. Deconvolution of the O 1s spectra indicates three main peaks around 531.08, 532.03, and 533.43 eV, which were assigned to C=O (oxygen doubly bonded to aromatic carbon) [10, 15], C–O (oxygen singly bonded to aliphatic carbon) [16, 17], and phenolic (oxygen singly bonded to aromatic carbon) [16, 17] groups, respectively, as shown in Fig. 2.2d. On the other hand, the pristine GO shows an additional peak at a higher binding energy (534.7 eV) [12], which originates from the chemisorbed/ intercalated adsorbed water molecules.

An important parameter that can be used to characterize the degree of oxidation in GO is the fraction of sp^2 carbon, which can be estimated by dividing the area under the sp^2 peak by the area of C 1s peak. In pristine GO, the fraction of sp^2 carbons is only ~40 %. During thermal reduction, the amount of sp^2 carbons gradually increases due to the loss of oxygen. It reaches a maximum value of ~80 % at

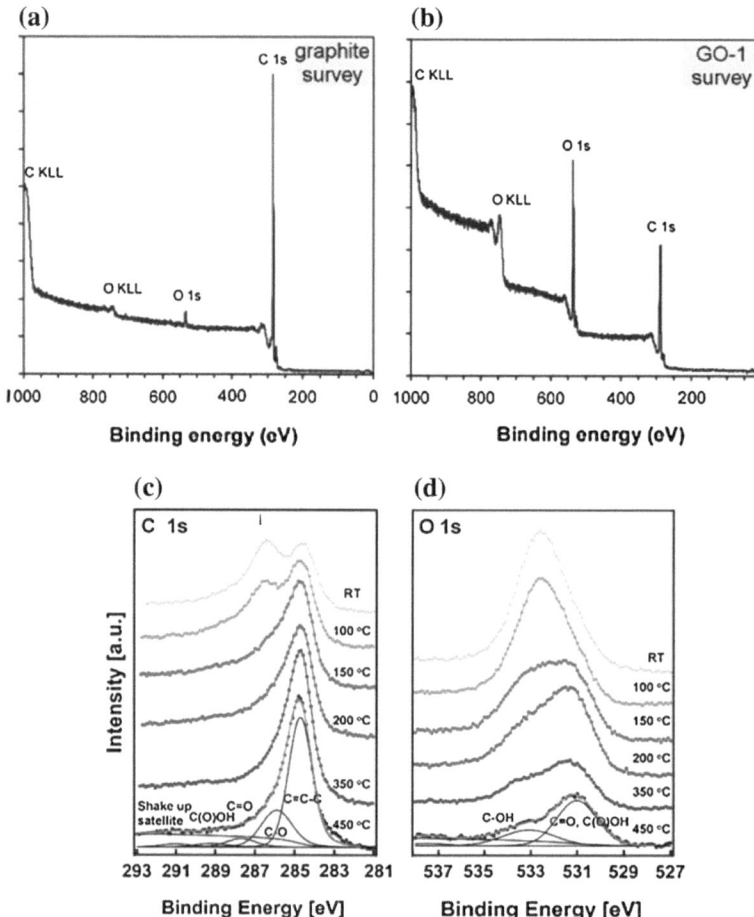

Fig. 2.2 a, b are XPS spectra of graphite and GO separately. Reprinted with permission from Ref. [4]. Copyright (2006) American Chemical Society. **c, d** are high-resolution XPS spectra of the C 1s and O 1s signals in GO. Reprinted with permission from Ref. [10]. Copyright (2009) WILEY-VCH Verlag GmbH & Co. KGaA, Weinheim

an oxygen content of ~8 atom% (C/O ratio 12.5:1) [10, 17]. This suggests that the remaining oxygen is responsible for ~20 % of the sp^3 bonding.

In addition, XPS can further elucidate change of the functional groups of GO at different degrees of oxidation [18]. At the initial stage, during increasing the oxidation level, intensity of the C–OH and –O–C=O functional groups increases and the corresponding intensity of C–C decreases. In the following, further increase of the oxidation results in the formation of epoxide groups along with the hydroxyl and carboxyl groups. Afterwards, continually oxidizing GO will enhance the intensity of epoxide groups but reduce the intensity of hydroxyl and carboxyl groups simultaneously.

2.1.3 XANES

XANES is another powerful tool for characterizing the GO materials. It provides information on degree of bond hybridization in mixed sp^2/sp^3 bonded carbon, specific bonding configurations of functional atoms, and degree of alignment of the graphitic crystal structures within GO [19]. As shown in Fig. 2.3a, the high-resolution C K-edge XANES spectrum of GO demonstrates distinctive unoccupied π^* and σ^* states around 285.2 and 293.03 eV, respectively, which can be primarily assigned to the 1s $\rightarrow$ π^* and 1s $\rightarrow$ σ^* transitions in the graphitic carbon atoms [13, 20]. Meanwhile, there is a broadening of the absorption peak at 289.3 eV, corresponding to the 1s $\rightarrow$ π^* transitions in the carbon atoms bonded with the oxygen atoms. The ratio of π^*/σ^* peaks at the C K-edge can be used to estimate the relative concentration of sp^2 domain configurations in an sp^3 matrix of GO, where carbon atoms are attached to oxygen groups. Thus, this ratio gives the degree of oxidation in GO [13].

On the other hand, the O K-edge XANES spectrum of GO [13, 21] generally shows several distinct absorption peaks at 531.5, 534.0, 535.5, 540.0, 542.0, and 544.5 eV, respectively, as presented in Fig. 2.3b. These peaks have been separately assigned to $\pi^*(C=O)$, $\pi^*(C–O)$, $\sigma^*(O–H)$, $\sigma^*(C–O)$, $\sigma^*(C=O)$, and $\sigma^*(C=O)$ [21]. Therefore, the O K-edge spectrum further helps to realize the GO structure, which clarifies the chemical composition of the oxygen-containing groups. According to Pacilé et al.'s measurements [21], the groups of epoxy, hydroxyl and carbonyl are likely to be attached to aromatic rings, while the carboxyl groups are likely to be bonded to the edges of the GO sheets.

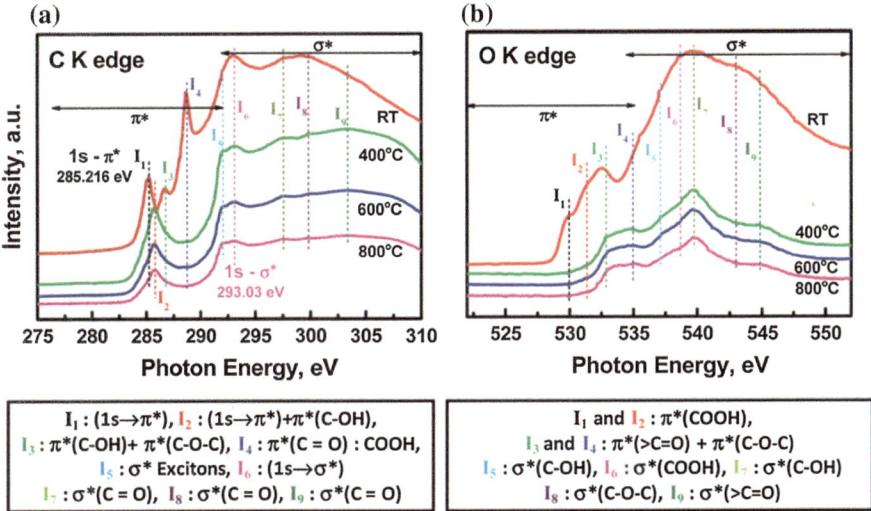

Fig. 2.3 High-resolution **a** C K-edge and **b** O K-edge synchrotron NEXAFS spectra at different reduction temperatures. The spectra were shifted in y-scale for clarity. Reprinted with permission from Ref. [13]. Copyright (2011) American Chemical Society

2.1.4 FT-IR

FT-IR spectroscopy is recognized as an important tool to study different types of functional groups. From the FT-IR characterization, Lee et al. [22] found that there are four main peaks in the FT-IR spectra of GO, as described detailedly in the following. Firstly, the peak at $1,050$ cm^{-1} arises from epoxide groups (C–O–C). Secondly, the one centered at $1,680$ cm^{-1} corresponds to the vibrational mode of the ketone groups (–C=O). Another peak at $1,380$ cm^{-1} is assigned to a C–O vibrational mode. Finally, the peak at $3,470$ cm^{-1} denotes C–OH stretching.

Bagri et al. [15] studied the structural evolution of GO during reduction. At the initial stage where GO is mildly annealed at 423 K, the FT-IR characterization shows that pristine GO was composed of hydroxyls ($3,050$–$3,800$ cm^{-1}), carbonyls ($1,750$–$1,850$ cm^{-1}), carboxyls ($1,650$–$1,750$ cm^{-1}), C=C ($1,500$–$1,600$ cm^{-1}) and ethers and/or epoxides ($1,000$–$1,280$ cm^{-1}), as displayed in part (1) of Fig. 2.4a. After annealing to 448 K for 5 min, the FT-IR spectrum changes, indicating that the carboxyl groups are removed, as shown in part (2) of Fig. 2.4a. When further annealing to $1,023$ K, hydroxyls disappear continuously and some ether groups are formed, as displayed in part (3) of Fig. 2.4a. In fact, the hydroxyls are not detected in infrared spectra at temperature above 773 K.

Similarly, Krishnamoorthy et al. [18] also investigated the structure of GO with different degrees of oxidation using FT-IR spectroscopy, as presented in Fig. 2.4b. The sp^2/sp^3 ratios of GO samples from S-1 to S-6 are 2.15, 1.52, 0.36, 0.31, 0.27, and 0.25, respectively. The FT-IR show band at $1,573$ cm^{-1} due to the presence of C–C stretching in graphitic domains of sample S-1. With further increase in oxidation level, the FT-IR spectrum reveal the presence of C=O ($1,720$ cm^{-1}), C–O ($1,050$ cm^{-1}), C–O–C ($1,250$ cm^{-1}), C–OH ($1,403$ cm^{-1}) in the GO samples. The peak at $1,620$ cm^{-1} is a resonance peak that can be assigned to the C–C stretching and absorbed hydroxyl groups in the GO.

2.1.5 Raman

Raman spectroscopy is a non-destructive technique that is widely used to obtain structural information of carbon materials. Usually, the Raman spectrum of a GO film displays a D band at $\sim 1,350$ cm^{-1} and a broad G band at $\sim 1,580$ cm^{-1} [11]. The G peak is the characteristic of all sp^2-hybridized carbon networks, which originates from the first-order scattering from the doubly degenerate E$_{2g}$ phonon modes of graphite in the Brillouin zone center as well as bond stretching of sp^2 carbon pairs in both rings and chains. Meanwhile, the D peak is due to the breathing mode of aromatic rings [23], which comes from the structural imperfections created by the attachment of oxygenated groups on the carbon basal plane [19]. Therefore, the D-peak intensity is often used as a measure for the degree of disorder [24]. Generally, the integrated intensity ratio of the D- and G-bands (I_D/I_G) indicates the oxidation degree and the size of sp^2 ring clusters in a sp^3/sp^2 hybrid network of carbon atoms. Another peak around $2,680$ cm^{-1}, usually called 2D peak, is the overtone

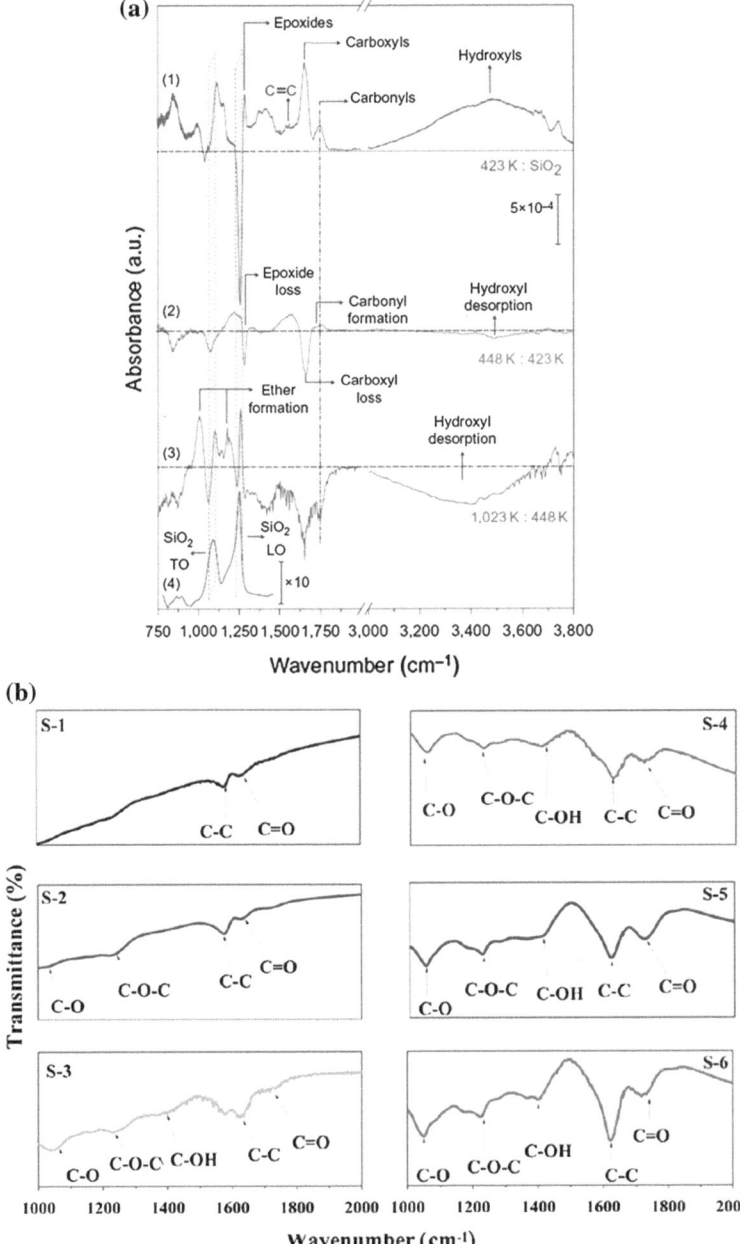

Fig. 2.4 a FT-IR spectra of single-layer GO: annealing at 423 K and referenced to the bare oxidized silicon substrate the spectrum (*1*); annealing to 448 K (*2*), referenced to spectrum (*1*); annealing to 1,023 K (*3*), referenced to spectrum (*2*); full SiO₂ absorption of the oxide referenced to H-terminated silica (*4*). Reprinted with permission from Ref. [15]. Copyright (2010) Nature Publishing Group. **b** FT-IR spectra of GO with different degrees of oxidation from samples S-1 to S-6. Reprinted with permission from Ref. [18]. Copyright (2012) Elsevier Ltd.

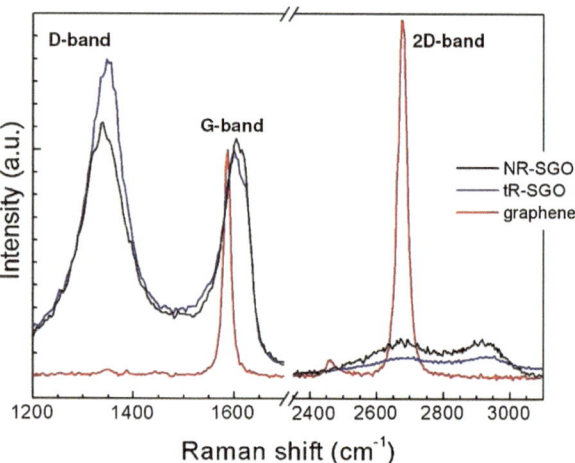

Fig. 2.5 Raman spectra of single-sheet GO (NR-SGO), single-sheet thermally reduced GO (tR-SGO), and mechanically exfoliated single-sheet graphene on SiO_2/Si substrates normalized to the G-peak intensity. Reprinted with permission from Ref. [28]. Copyright (2008) American Chemical Society

of the D peak, which reflects the number of graphene layers [25, 26]. The 2D peak is attributed to double resonance transitions resulting in production of two phonons with opposite momentum. Different from the D peak, which is only Raman active in the presence of defects, the 2D peak is active even in the absence of any defects [27].

Figure 2.5 compares typical Raman spectra of GO, RGO and graphene, which were recorded at an excitation wavelength of 532 nm [28]. Generally, the Raman spectra of GO and RGO are very close to each other, both exhibiting the same D, G and 2D peak positions. The only distinct difference is the I_D/I_G ratio since RGO usually has much lower oxidation degree than that of GO. Compared with graphene, the prominent D peak along with weak and broad 2D peak are the main signs of GO due to structural disorder [29]. Usually, the D peak is absent in graphene [30]. Meanwhile, the mechanically exfoliated graphene has a much strong and sharp 2D peak.

Krishnamoorthy et al. [18] studied the Raman spectra of GO with different degrees of oxidation. During oxidation of graphite, the G band is shifted from $1,570 \text{ cm}^{-1}$ (G band of graphite) towards a higher wavenumber ($1,585 \text{ cm}^{-1}$) and the D band has higher intensity, which can attributed to the formation of defects and disorder such as the presence of in-plane heteroatoms, grain boundaries, aliphatic chains, and so on. On the other hand, the intensity of the 2D band is diminished after oxidation. Moreover, a new band appears around $2,950 \text{ cm}^{-1}$, which is denoted as D + G band. The reduction in the intensity of the 2D band is attributed to breaking of stacking order associated with oxidation reaction.

2.2 Microscopic Characterization

2.2.1 TEM

TEM is a common microscopic technique to feature the atomic structures of nanomaterials. By means of TEM, one can directly image the lattice atoms and

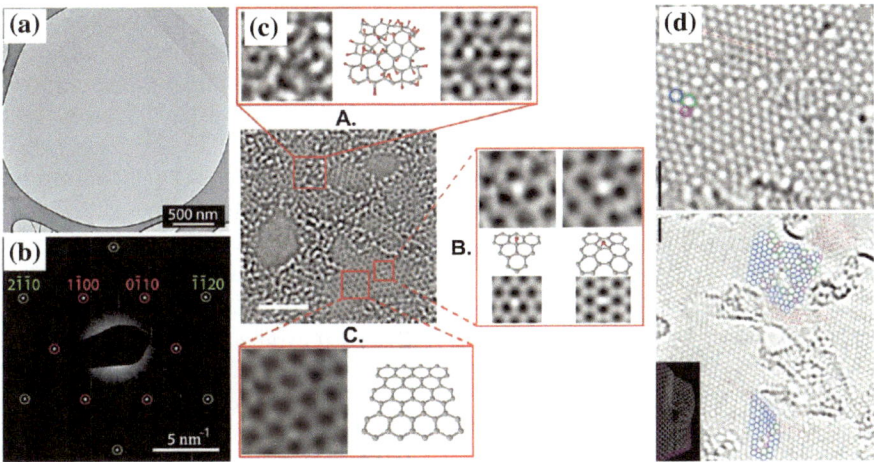

Fig. 2.6 a TEM image of a single GO sheet on a lacey carbon support and **b** its SAED of the center of the region. Reprinted with permission from Ref. [34]. Copyright (2009) American Chemical Society. **c** Aberration-corrected TEM image of a single sheet of suspended GO with a scale bar of 2 nm. Expansion (*A*) shows, from *left* to *right*, a 1 nm^2 enlarged oxidized region of the GO, then a proposed possible atomic structure of this region with carbon atoms in *gray* and oxygen atoms in *red*, and finally the average of a simulated TEM image of the proposed structure. Expansion (*B*) focuses on the *white spot* on the graphitic region, which moves along the graphitic region. Expansion (*C*) shows a 1 nm^2 graphitic portion from the planewave reconstruction of a focal series of GO and the atomic structure of this region. Reprinted with permission from Ref. [32]. Copyright (2010) WILEY-VCH Verlag GmbH & Co. KGaA, Weinheim. **d** TEM images of extended topological defects and deformations in RGO, including pentagons, heptagons, distortions and strain in the surrounding lattice. Carbon pentagons, hexagons, and heptagons are indicated in *magenta*, *blue*, and *green*, respectively. The *red dashed lines* denote directions with strong deformations in the lattice. A relaxed structural model similar to the observed configuration is shown in *inset*. Reprinted with permission from Ref. [31]. Copyright (2010) American Chemical Society

topological defects in GO [21, 31–33], which is of significant importance to explore their atomic structure. A typical TEM image of a GO monolayer shown in Fig. 2.6a indicates it is highly electron transparent even in comparison to the thin-film carbon support [34]. A selected area electron diffraction (SAED) pattern from the monolayer region of the GO film is shown in Fig. 2.6b. It should be noticed that clear diffraction spots are observed, indicating that the crystalline order of the original graphene lattice is preserved. Meanwhile, for GO samples with different oxidation degrees, their TEM images show different transparencies because of different number of layers in the stacked structure of GO [18]. At lower oxidation degree, the samples contain less oxygenated functional groups which limits them in terms of exfoliation into monolayers or few layers after the exfoliation process. With increasing the oxidation level, GO samples become highly transparent since the samples possess high amounts of oxygenated functional groups and can be easily exfoliated into monolayers or just a few layers of GO after ultrasonic treatment.

High-resolution TEM (HRTEM) is able to directly image the honeycomb lattice along with structural disorder in GO. Using HRTEM, Erickson et al. [32] demonstrated that the specific atomistic features of GO shows three major regions, which are holes, graphitic regions, and highly contrast disordered regions with approximate area percentages of 2, 16, and 82 %, respectively (Fig. 2.6c). It is proposed that the holes in GO are formed by releasing CO and CO_2 during the aggressive oxidation and sheet exfoliation. The graphitic regions are resulted from incomplete oxidation of the basal plane, which preserves the honeycomb structure of graphene. Meanwhile, the disordered regions of the basal plane are originated from abundant oxygen-containing groups aggregated in these regions, including hydroxyls, epoxies, and carbonyls. Gómez-Navarro et al. [31] further unraveled the topological defects in GO using aberration-corrected HRTEM. They pointed out the dominant clustered pentagons and heptagons, as well as the existence of in-plane distortions and strain in the surrounding lattice of GO, as displayed in Fig. 2.6d.

2.2.2 STM

Another useful microscopic technique is STM. Previously, several groups have utilized STM to study the surface of GO and observed highly defective regions [35–39]. According to Gómez-Navarro et al.'s measurement [35], pristine graphene and oxidized regions are distinguishable through the bright spots as shown in Fig. 2.7a, where the oxidized regions are marked by green contours. By estimating the ratio of oxidized regions, the degree of functionalization can be obtained. Then, Kudin et al. [36] compared the STM images of highly oriented pyrolytic graphite (HOPG) and GO, as depicted in Fig. 2.7b. The STM image of HOPG is presented in the inset at the left bottom of Fig. 2.7b, which is in a highly crystalline order. In contrast, the STM image of GO appears rough, featuring a peak-to-peak topography of 1 nm. This roughness is caused by functional groups and defects. Fourier transformation of the STM image (inset at the right top of Fig. 2.7b) shows a clear signature of a graphitic backbone where the hexagonal symmetry is highlighted by manually added lines. This indicates reemergence of graphitic order during the reduction process.

Pandey et al. [38] also examined the oxidized regions of GO and surprisingly observed a periodic arrangement of oxygen atoms, which spanned over a few nanometers, as shown in Fig. 2.7c. This periodic arrangement can be understood by a structural model illustrated in Fig. 2.7d, where oxygen atoms are arranged in a rectangular lattice, suggesting a series of epoxy groups are presented in strips.

In addition, through the STM images, Doğan et al. [39] shows the defects of vacancies and adatoms in the electrochemically reduced GO. Particularly, from the atomically resolved STM images obtained at different parts of the same reduced GO sample, they observed a moiré pattern, where the GO sample exhibits

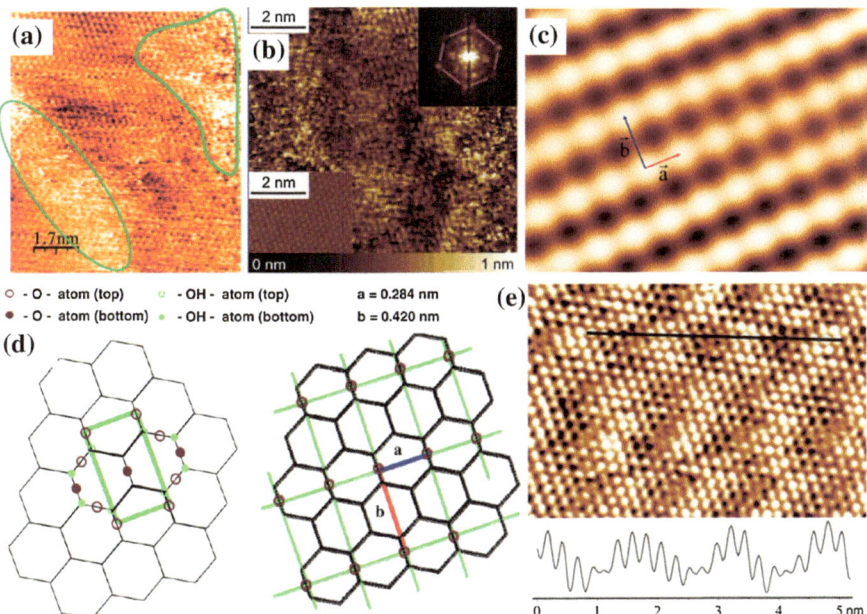

Fig. 2.7 a STM image of a GO monolayer on a HOPG substrate, taken under ambient conditions. Oxidized regions are marked by *green* contours. Reprinted with permission from Ref. [35]. Copyright (2007) American Chemical Society. **b** STM image of a GO monolayer on a HOPG substrate and its Fourier transform (*inset at top right*). Reprinted with permission from Ref. [36]. Copyright (2008) American Chemical Society. **c** High-resolution STM image of the oxidized regions of GO revealing a rectangular lattice of oxygen atoms. **d** A structural model of GO to show the case of (**c**). Reprinted with permission from Ref. [38]. Copyright (2008) Elsevier B.V. **e** An atomically resolved STM image of the electrochemically reduced GO film showing a Moiré pattern with a periodicity of around 1.45 nm, where the lower part of the figure shows the height profiles taken along the *black line*. Reprinted with permission from Ref. [39]. Copyright (2013) Elsevier B.V.

a hexagonal lattice with an average periodicity of about ~1.45 nm, as presented in Fig. 2.7e. This hexagonal moiré structures originate from the lattice mismatch between the graphene and the surface of hexagonally close-packed (hcp) Au solid, which leads to incommensurate structures.

2.2.3 Other Microscopic Tools and Combined Techniques

Besides TEM and STM, other microscopic means are also helpful to reveal the atomistic structures of GO. For example, AFM directly gives the apparent thickness as well as the number of layers of GO [6, 35, 37, 38]. Generally, analysis of a larger number of AFM images reveals that GO sheets have lateral dimensions

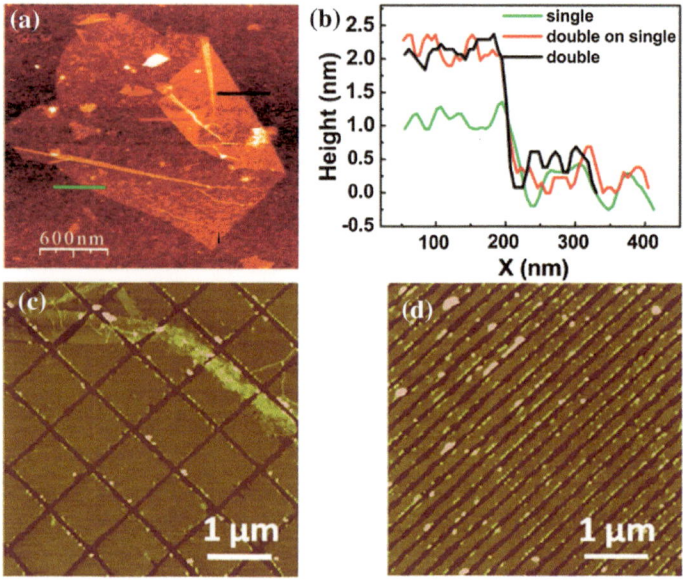

Fig. 2.8 a AFM image of a GO monolayer deposited on a SiO_2 substrate, showing a back-folded edge; **b** AFM section profiles along the three different lines in panel (**a**), revealing mono-, bi-, and tri-layer structures, respectively. Reprinted with permission from Ref. [35]. Copyright (2007) American Chemical Society. AFM nanolithography of GO sample into square arrays (**c**) and linear arrays (**d**). Reprinted with permission from Ref. [41]. Copyright (2010) AIP Publishing LLC

of 100–5,000 nm and heights in the range of 1.1–15 nm. Approximately 80 % of the GO sheets displayed a height of 1.1 ± 0.2 nm. Figure 2.8a displays an AFM image of a GO sheet whose upper right edge is double-folded onto itself, as can be concluded from the cross-sectional profiles depicted in Fig. 2.8b. Meanwhile, AFM can be utilized for nanolithography [40–42], where GO samples can be etched into different shapes, such as squares, ribbons, quantum dots, as shown in Fig. 2.8c–d.

In addition, using high-resolution annular dark field (ADF) imaging in a STEM instrument, Mkhoyan et al. [43] investigated the oxygen distribution on a GO monolayer. Their results indicated that the degree of oxidation fluctuates at nanometer scale, suggesting the presence of sp^2 and sp^3 carbon clusters of a few nanometers. Employing STEM combined with electron energy loss spectroscopy (EELS), they were then able to measure the fine structure of the carbon, oxygen K-edges, and low-loss electronic excitations in GO [43]. They found that the oxygen atoms are randomly attached to the graphene sites, converting the sp^2 carbon in graphene to sp^3 hybridization.

Also employing the STEM, Zhu et al. [44] analyzed structural changes of GO nanoribbons during thermal annealing, as shown in Fig. 2.9. Usually, the chemical changes of GO nanoribbons are directly related to the planarity and the sp^2-carbon

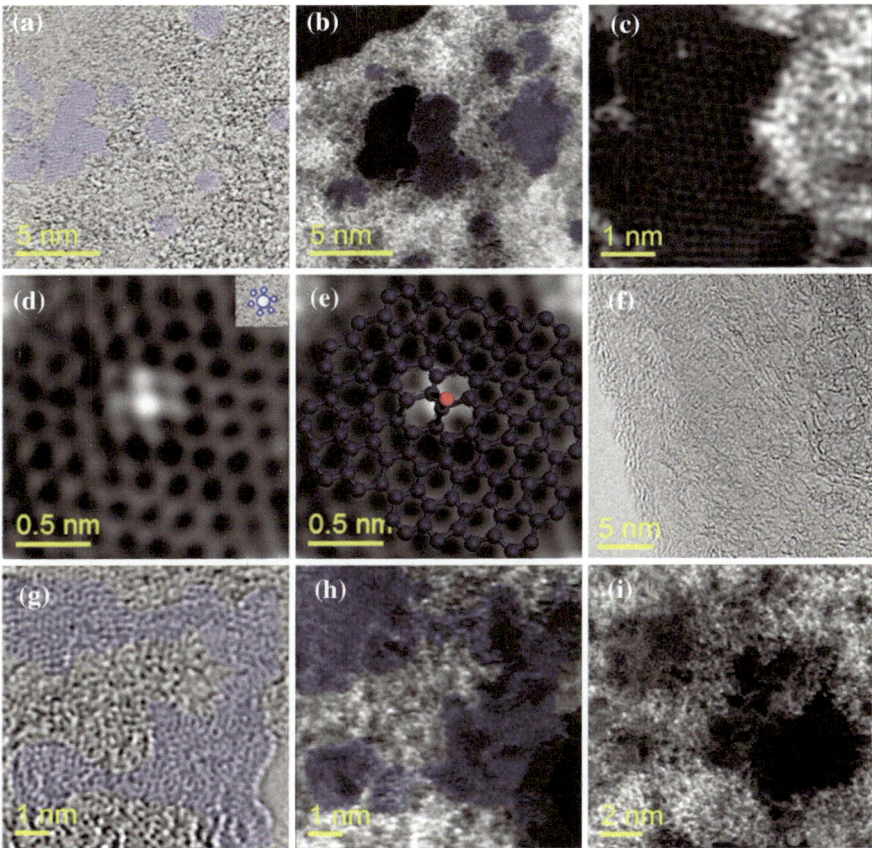

Fig. 2.9 STEM images of the non-annealed GO nanoribbons (**a–e**) and GO nanoribbons annealed at 200 °C for 10 min (**f–i**). **a** Bright field (BF) and **b** high-angle annular dark-field (HAADF) STEM image of GO nanoribbons with graphene regions shown in *blue*. **c** HAADF STEM image of a monolayer graphene region (same area as the center in **b**). **d** High-resolution image of monolayer graphene with an oxygen functional group on the basal surface after applying a filter of the raw HAADF STEM image. **e** HAADF STEM image of the same region shown in (**d**) with an overlay structure sketch. **f** BF STEM image of annealed GO nanoribbons with stacked graphene layers. **g** BF STEM image and **h** HAADF STEM image of annealed GO nanoribbons with graphene regions shown in *blue*. **i** HAADF STEM image of annealed GONRs with holes. Reprinted with permission from Ref. [44]. Copyright (2012) American Chemical Society

structures. As shown in Fig. 2.9a–e, in the non-annealed samples, nanoscale regions of monolayer graphene are observed, where the region sizes typically range from 1 to 2 nm. But in some directions it can be as long as 3–5 nm, as shown in Fig. 2.9a, b. However, the majority of the film is functionalized with oxygen-containing groups, as indicated in Fig. 2.9d, e. Meanwhile, the graphitic regions are small and isolated in a non-annealed sample. During thermal annealing, the structure of GO sample changes, even forming some holes, as indicated by the black regions in Fig. 2.9h, i.

References

1. He, H., Riedl, T., Lerf, A., Klinowski, J.: J. Phys. Chem. **100**, 19954–19958 (1996)
2. Lerf, A., He, H., Riedl, T., Forster, M., Klinowski, J.: Solid State Ion. **101–103**, 857–862 (1997)
3. Lerf, A., He, H., Forster, M., Klinowski, J.: J. Phys. Chem. B **102**, 4477–4482 (1998)
4. Szabó, T., Berkesi, O., Forgó, P., Josepovits, K., Sanakis, Y., Petridis, D., Dékány, I.: Chem. Mater. **18**, 2740–2749 (2006)
5. Cai, W.W., Piner, R.D., Stadermann, F.J., Park, S., Shaibat, M.A., Ishii, Y., Yang, D.X., Velamakanni, A., An, S.J., Stoller, M., An, J.H., Chen, D.M., Ruoff, R.S.: Science **321**, 1815–1817 (2008)
6. Stankovich, S., Dikin, D.A., Piner, R.D., Kohlhaas, K.A., Kleinhammes, A., Jia, Y., Wu, Y., Nguyen, S.T., Ruoff, R.S.: Carbon **45**, 1558–1565 (2007)
7. Gao, W., Alemany, L.B., Ci, L.J., Ajayan, P.M.: Nat. Chem. **1**, 403–408 (2009)
8. Zhang, Q., Scrafford, K., Li, M., Cao, Z., Xia, Z., Ajayan, P.M., Wei, B.: Nano Lett. **14**, 1938–1943 (2014)
9. Casabianca, L.B., Shaibat, M.A., Cai, W.W., Park, S., Piner, R., Ruoff, R.S., Ishii, Y.: J. Am. Chem. Soc. **132**, 5672–5676 (2010)
10. Mattevi, C., Eda, G., Agnoli, S., Miller, S., Mkhoyan, K.A., Celik, O., Mastrogiovanni, D., Granozzi, G., Garfunkel, E., Chhowalla, M.: Adv. Funct. Mater. **19**, 2577–2583 (2009)
11. Yang, D., Velamakanni, A., Bozoklu, G., Park, S., Stoller, M., Piner, R.D., Stankovich, S., Jung, I., Field, D.A., Ventrice, C.A., Ruoff, R.S.: Carbon **47**, 145–152 (2009)
12. Akhavan, O.: Carbon **48**, 509–519 (2010)
13. Ganguly, A., Sharma, S., Papakonstantinou, P., Hamilton, J.: J. Phys. Chem. C **115**, 17009–17019 (2011)
14. Lahaye, R.J.W.E., Jeong, H.K., Park, C.Y., Lee, Y.H.: Phys. Rev. B **79**, 125435 (2009)
15. Bagri, A., Mattevi, C., Acik, M., Chabal, Y.J., Chhowalla, M., Shenoy, V.B.: Nat. Chem. **2**, 581–587 (2010)
16. Hontoria-Lucas, C., López-Peinado, A.J., López-González, JdD, Rojas-Cervantes, M.L., Martín-Aranda, R.M.: Carbon **33**, 1585–1592 (1995)
17. Schniepp, H.C., Li, J.L., McAllister, M.J., Sai, H., Herrera-Alonso, M., Adamson, D.H., Prud'homme, R.K., Car, R., Saville, D.A., Aksay, I.A.: J. Phys. Chem. B **110**, 8535–8539 (2006)
18. Krishnamoorthy, K., Veerapandian, M., Yun, K., Kim, S.J.: Carbon **53**, 38–49 (2013)
19. Chen, D., Feng, H., Li, J.: Chem. Rev. **112**, 6027–6053 (2012)
20. Saxena, S., Tyson, T.A., Negusse, E.: J. Phys. Chem. Lett. **1**, 3433–3437 (2010)
21. Pacilé, D., Meyer, J.C., Fraile Rodríguez, A., Papagno, M., Gómez-Navarro, C., Sundaram, R.S., Burghard, M., Kern, K., Carbone, C., Kaiser, U.: Carbon **49**, 966–972 (2011)
22. Lee, D.W., De Los Santos, V.L., Seo, J.W., Felix, L.L., Bustamante, D.A., Cole, J.M., Barnes, C.H.W.: J. Phys. Chem. B **114**, 5723–5728 (2010)
23. Tuinstra, F., Koenig, J.L.: J. Chem. Phys. **53**, 1126–1130 (2003)
24. Ferrari, A.C., Robertson, J.: Phys. Rev. B **61**, 14095 (2000)
25. Ferrari, A.C., Meyer, J.C., Scardaci, V., Casiraghi, C., Lazzeri, M., Mauri, F., Piscanec, S., Jiang, D., Novoselov, K.S., Roth, S.: Phys. Rev. Lett. **97**, 187401 (2006)
26. Gupta, A., Chen, G., Joshi, P., Tadigadapa, S., Eklund, P.C.: Nano Lett. **6**, 2667–2673 (2006)
27. Eda, G., Chhowalla, M.: Adv. Mater. **22**, 2392–2415 (2010)
28. Jung, I., Dikin, D.A., Piner, R.D., Ruoff, R.S.: Nano Lett. **8**, 4283–4287 (2008)
29. Robinson, J.A., Wetherington, M., Tedesco, J.L., Campbell, P.M., Weng, X., Stitt, J., Fanton, M.A., Frantz, E., Snyder, D., VanMil, B.L.: Nano Lett. **9**, 2873–2876 (2009)
30. Malard, L.M., Pimenta, M.A., Dresselhaus, G., Dresselhaus, M.S.: Phys. Rep. **473**, 51–87 (2009)
31. Gómez-Navarro, C., Meyer, J.C., Sundaram, R.S., Chuvilin, A., Kurasch, S., Burghard, M., Kern, K., Kaiser, U.: Nano Lett. **10**, 1144–1148 (2010)

32. Erickson, K., Erni, R., Lee, Z., Alem, N., Gannett, W., Zettl, A.: Adv. Mater. **22**, 4467–4472 (2010)
33. Xie, J., Tu, F., Su, Q., Du, G., Zhang, S., Zhu, T., Cao, G., Zhao, X.: Nano Energy **5**, 122–131 (2014)
34. Wilson, N.R., Pandey, P.A., Beanland, R., Young, R.J., Kinloch, I.A., Gong, L., Liu, Z., Suenaga, K., Rourke, J.P., York, S.J.: ACS Nano **3**, 2547–2556 (2009)
35. Gómez-Navarro, C., Weitz, R.T., Bittner, A.M., Scolari, M., Mews, A., Burghard, M., Kern, K.: Nano Lett. **7**, 3499–3503 (2007)
36. Kudin, K.N., Ozbas, B., Schniepp, H.C., Prud'homme, R.K., Aksay, I.A., Car, R.: Nano Lett. **8**, 36–41 (2008)
37. Paredes, J.I., Villar-Rodil, S., Solis-Fernandez, P., Martinez-Alonso, A., Tascon, J.M.D.: Langmuir **25**, 5957–5968 (2009)
38. Pandey, D., Reifenberger, R., Piner, R.: Surf. Sci. **602**, 1607–1613 (2008)
39. Doğan, H.Ö., Ekinci, D., Demir, Ü.: Surf. Sci. **611**, 54–59 (2013)
40. Masubuchi, S., Ono, M., Yoshida, K., Hirakawa, K., Machida, T.: Appl. Phys. Lett. **94**, 082107 (2009)
41. He, Y., Dong, H., Li, T., Wang, C., Shao, W., Zhang, Y., Jiang, L., Hu, W.: Appl. Phys. Lett. **97**, 133301 (2010)
42. Lu, G., Zhou, X., Li, H., Yin, Z., Li, B., Huang, L., Boey, F., Zhang, H.: Langmuir **26**, 6164–6166 (2010)
43. Mkhoyan, K.A., Contryman, A.W., Silcox, J., Stewart, D.A., Eda, G., Mattevi, C., Miller, S., Chhowalla, M.: Nano Lett. **9**, 1058–1063 (2009)
44. Zhu, Y., Li, X., Cai, Q., Sun, Z., Casillas, G., Jose-Yacaman, M., Verduzco, R., Tour, J.M.: J. Am. Chem. Soc. **134**, 11774–11780 (2012)

Chapter 3
Structural Modeling and Physical Properties

Abstract The structural modeling of GO is helpful to establish an atomistic understanding of GO. From the structural models, one can further explain the various physical and chemical properties of GO observed in experiments or exploit its potential applications in various fields. So far, numerous computational simulations have been carried out to explore the structure of GO, investigating effects of coverage, ratio and arrangement of the oxygen-containing groups. Generally, a structural model with epoxides and hydroxyls orderly arranged in a chained form on the graphene basal plane is considered to be thermodynamic stable for GO. On the other hand, the fundamental physical properties of GO are closely related to the coverage, ratio and arrangement of the functional groups. The electronic property of GO is tunable, mainly depending on the coverage and ratio of the epoxides and hydroxyls. The tunable band gap also leads to the tunable optical property of GO. Moreover, the mechanical property of GO mainly relies on the coverage and arrangement the functional groups as well as the thickness. Finally, composing with other nanomaterials is an effective way to further tailor the mechanical and optical properties of GO.

3.1 Structural Modeling

3.1.1 Sketching Models

Early in 1939, Hofmann and Holst proposed a structural model of GO with only epoxide groups [1]. They supposed that the oxygen was bound to the carbon atoms of the hexagon layer planes by epoxide linkages with an ideal formula of C_2O, as shown in Fig. 3.1a. Then in 1947, considering the hydrogen content of GO, Ruess suggested a structural model incorporated with hydroxyl groups [2], as shown in Fig. 3.1b. This model also indicates that the basal plane structure of GO is in an sp^3 hybridization form, rather than the sp^2 hybridized model of Hofmann and Holst [1]. Scholz and Boehm reconsidered the stoichiometric ratio and revised

© The Author(s) 2015
J. Zhao et al., *Graphene Oxide: Physics and Applications*,
SpringerBriefs in Physics, DOI 10.1007/978-3-662-44829-8_3

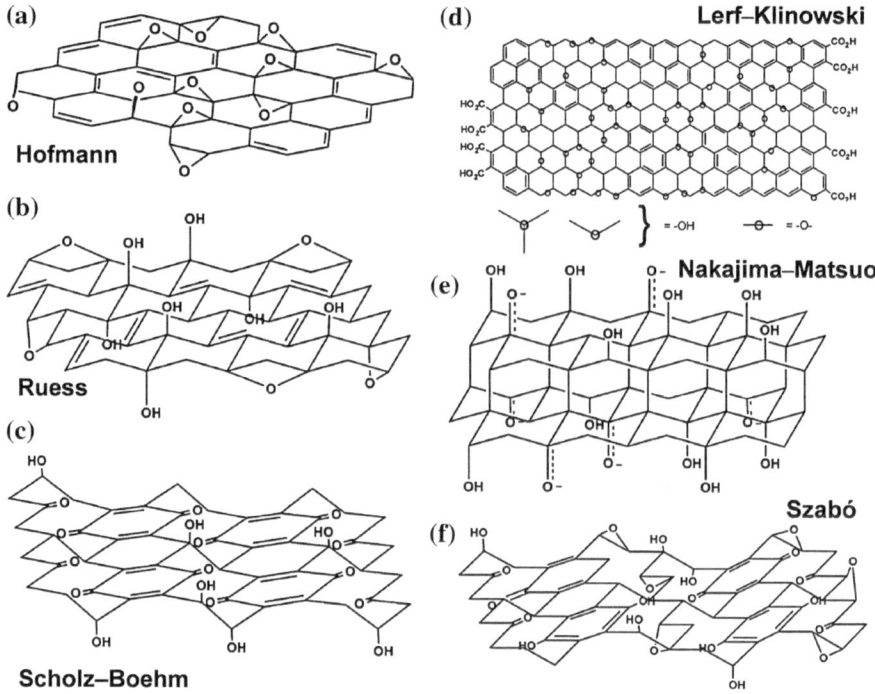

Fig. 3.1 Sketching models proposed for GO. Reprinted with permission from Ref. [9]. Copyright (2006) American Chemical Society

Ruess's model, suggesting a model that consisted of ribbons of conjugated carbon backbone and regular quinoidal species but completely removed the epoxide groups [3], as shown in Fig. 3.1c.

However, all these early models are generally based on the chemical formula and elemental analysis; few spectrum information are featured to support these models. Therefore, on the basis of the ^{1}H and ^{13}C solid-state NMR spectra, Lerf and coworkers [4–6] replaced the previous models by a structural model with randomly distributed flat aromatic and wrinkled regions. The flat aromatic regions consist of unoxidized benzene rings and the wrinkled regions beared C=C, C–OH, and ether groups, as shown in Fig. 3.1d. On the other hand, starting from the XRD information, Nakajima et al. [7, 8] proposed a poly-like $(C_2F)_n$ model by fluorination of GO, as shown in Fig. 3.1e. In this model, two carbon layers link to each other by sp^3 C–C bonds perpendicular to the layers, where carbonyl and hydroxyl groups are present in relative amounts depending on the level of hydration. Particularly, based on the spectroscopic characterization of GO including FT-IR, XPS and NMR, Szabó et al. [9] revived the model of proposed by Scholz and Boehm and gave a model with a corrugated carbon network including a ribbon-like arrangement of flat carbon hexagons connected by C=C double bonds. Therefore, the resulting carbon skeleton is a mixture of the Ruess and

Scholz-Boehm skeleton, including a random distribution of two kinds of domains: the trans linked cyclohexane chairs and the corrugated hexagon ribbons [9], as shown in Fig. 3.1f. Some recent reviews [10–14] and book chapters [15] have described these sketching models of GO structures in further detail.

3.1.2 Atomistic Simulations

As mentioned in Chap. 2, experimental characterizations with various spectroscopic techniques have demonstrated that epoxide and hydroxyl groups are the two major functional groups on the basal plane of GO [16, 17]. Based on this conclusion, atomistic simulations were carried out to study the structure of GO.

In 2006, Schniepp et al. [18] investigated the thickness of GO using first-principles calculations. They found that an epoxide group is 1.9 Å above the carbon grid and a hydroxyl group is 2.2 Å above the carbon grid. Meanwhile, the protuberant carbon atoms are 0.3 and 0.7 Å above the carbon grid for epoxide and hydroxyl functional groups, respectively. When both epoxide and hydroxyl groups are placed on the carbon grid, a single-layered GO has a thickness of ~7.8 Å. Afterwards, employing DFT calculations, Boukhvalov and Katsnelson [19] studied three kinds of GO models: one with epoxides only (Fig. 3.2a), another with hydroxyls only (Fig. 3.2b), and the last one with both epoxide and hydroxyl groups (Fig. 3.2c). They found that

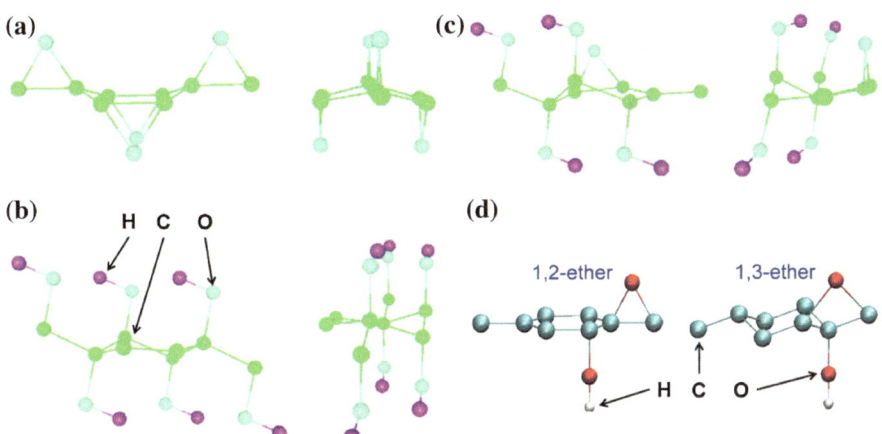

Fig. 3.2 The most stable configurations of GO with epoxide groups only (**a**), hydroxyl groups only (**b**), and both epoxide and hydroxyl groups (**c**). Carbon, oxygen, and hydrogen atoms are shown in *green*, *blue*, and *violet*, respectively. Reprinted with permission from Ref. [19]. Copyright (2008) American Chemical Society. **d** Schematic to show configurations of 1,2-ether and 1,3-ether epoxide groups, respectively. Reprinted with permission from Ref. [20]. Copyright (2009) American Physical Society

the oxygen-containing groups prefer to reside on both sides of graphene, while the hydroxyls are energetically favorable to sit at neighboring carbon atoms from opposite sides of the graphene. Moreover, GO with both epoxide and hydroxyl groups is more stable than the one with only epoxide or hydroxyl groups.

Later, Lahaye et al. [20] considered the epoxide in detail. They pointed out that the epoxide should be the 1,2-ether oxygen, while the 1,3-ether oxygen in GO is not energetically to be formed, as shown in Fig. 3.2d. They also proposed a GO structure with the 1,2-ether oxygen dominated and closely arranged. But at the reverse side of the carbon plane, the hydroxyl molecule is located. This arrangement repeats along the carbon network with subtle variations, leading to a random pattern when the oxidation over a macroscopic region appears.

Yan et al. [21, 22] studied the arrangement of epoxide and hydroxyl groups on graphene by the first-principles calculations. They found an energetically favorable configuration where the neighboring 1,2-hydroxyl pairs form hydroxyl chains via hydrogen-bond interaction and the epoxides are grouped next to these hydroxyl chains, as shown in Fig. 3.3a. The epoxide and hydroxyl units randomly deposited on the surfaces are expected to arrange themselves locally following these preferred patterns, forming specific oxygen-containing group chains with sp^2 carbon regions in between, as shown in Fig. 3.3b. This model was further confirmed by other groups. Wang et al. [23, 24] proposed a series of motifs for GO structures with epoxides only, hydroxyls only, and both epoxide and hydroxyl groups. For searching the energetically favorable motifs, four major factors have to be considered: (1) the accommodation of hydrogen bonding within an (–OH, –OH) pair or an (–OH, –O–) pair, (2) the reduction of strain and Coulomb repulsion among negatively charged oxygen ions, (3) the elimination of dangling bonds, and (4) the proximity of –OH to –O– group [23]. Based on these rules, four stable zero-dimensional (0D) motifs were obtained through the density functional theory (DFT) calculations [24], as shown in Fig. 3.3c. The hydroxyl group prefers to form a dimer with one hydroxyl locating next to the other below and above the carbon grid. Hydroxyl monomer is unstable because it creates a dangling bond that can be easily passivated by another hydroxyl. Similarly, the epoxide dimer forms the same pattern as that of the hydroxyl dimer to lower the strain. For GO with both epoxide and hydroxyl groups, these groups tend to aggregate together to form hydrogen bonds. Using these 0D motifs to build the structural model, GO structure with hydroxyl and epoxide chains siting closely is obtained, in agreement with the model proposed by Yan et al. [21, 22].

Moreover, by simulating XPS spectra, Zhang et al. [25] pointed out that epoxide and hydroxy groups can be closely packed together to form epoxides and epoxide-hydroxyl pairs in highly oxidized samples. Taking both the thermodynamic and kinetic factors into account, Lu et al. [26] also suggested that the hydroxyl chain is a very stable structure. However, the calculated vibrational frequencies and also previously simulated chemical shifts of this structure are not observed in infrared and NMR experiments. Therefore, kinetic effects during the synthesis of GO must play an important role in the GO structure. In addition, using the genetic algorithm combined with DFT calculations, Xiang et al. [27] performed global search of the lowest-energy structure of GO. It was indicated

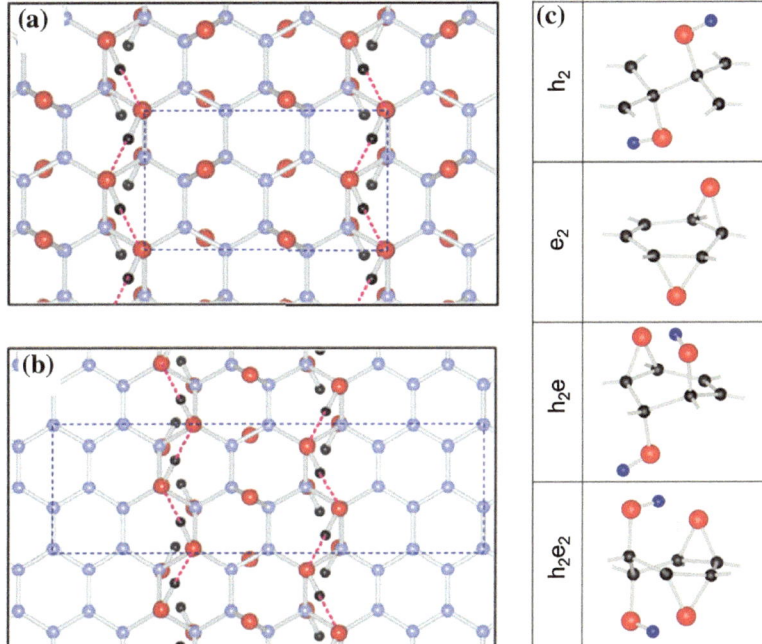

Fig. 3.3 a Fully-oxidized phase of GO ($C_6O_4H_2$) with epoxide and hydroxyls chains packed together; **b** GO structure ($C_{24}O_{10}H_8$) with hydroxyl-epoxide strips separated by sp^2 carbon. C, O and H atoms are represented by *grey*, *red* and *black spheres* individually and the *dashed rectangles* indicate the respective unit cells. The hydrogen bonds in the hydroxyl chains above the plane are indicated by *dashed lines*. Reprinted with permission from Ref. [21]. Copyright (2009) American Physical Society. **c** Stable motifs of the functional groups on GO: hydroxyl dimer (h_2), epoxide dimer (e_2), two-hydroxyl and one-epoxide trimer (h_2e), and two-hydroxyl and two-epoxy tetramer (h_2e_2). *Red*, *black*, and *blue balls* represent O, C, and H atoms, respectively. Reprinted with permission from Ref. [24]. Copyright (2010) American Physical Society

that the phase separation between bare graphene and fully oxidized graphene is thermodynamically favorable in GO, also consistent with Yan's reports [21, 22].

Furthermore, phase diagram of GO under different chemical potentials has been predicted. Starting from the formation energy, Yan et al. [21, 22] considered the relative stability of phases with different coverages and drew a ternary system. The formation energy (ΔE) is described by the Eq. (3.1):

$$\Delta E[x,y] = E[C_{1+x+y}O_x(OH)_{2y}] - (1 - x - y)E[C^*] - xE[C_2O] - yE[C_2(OH)_2],$$
(3.1)

where $E[Z]$ is the energy of a periodic phase Z. Each periodic phase can be specified by the relative amount of "free" sp^2 C atoms (denoted by C^*, corresponding to the C atoms not bonded to O), epoxide (C_2O), and the 1,2-hydroxyl pair $[C_2(OH)_2]$. The representative stoichiometry is $C^*_{1-x-y}(C_2O)_x[C_2(OH)_2]_y$, or equivalently $C_{1+x+y}O_x(OH)_{2y}$, with $0 \leq x \leq 1$, $0 \leq y \leq 1$, and $0 \leq x + y \leq 1$. The ordered

phases from their calculation are marked on a ternary diagram in Fig. 3.4a, where the
dashed lines indicate phases with the same ratio of epoxide versus the hydroxyl pair.

Wang et al. [24] also systematically investigated the structure of GO and constructed a structural phase diagram of the GO with respect to the chemical potentials of oxygen and hydrogen, as shown in Fig. 3.4b. The formation energy (ΔH_f) of
GO can be regarded as a function of the chemical potentials of O (μ_O) and H (μ_H):

$$\Delta H_f = E_{tot}(C_lO_mH_n) - E_{tot}(C_l) - m\mu_O - n\mu_H, \tag{3.2}$$

where $E_{tot}(C_lO_mH_n)$ and $E_{tot}(C_l)$ are the total energies of the unit cells of GO and
pristine graphene that contains l carbon, m oxygen, and n hydrogen atoms, respectively. The reference energies for μ_O and μ_H are taken as the total energies of gas-
eous O_2 and H_2 molecules, respectively. Fully covered GO phases without any sp^2
carbon, such as the hydroxyl-only, epoxide-only, and mixed hydroxyl and epoxide
ones, are thermodynamically stable, but only exist under stringent experimental
conditions due to competition with the formation of water. Considering the kinetic
factors, GO with both functional groups and sp^2 carbons is a kinetically hindered
metastable phase, even though it is routinely observed in experiments.

Since experimental characterizations indicate that GO is amorphous, there have
also been some studies on the amorphous structural model of GO. Using Monte Carlo
(MC) method, Paci et al. [28] randomly placed the epoxide and hydroxyl groups on
either side of a graphene basal plane. They found that hydrogen bonds are formed
between the functional groups, such as hydroxyl-hydroxyl and hydroxyl-epoxide

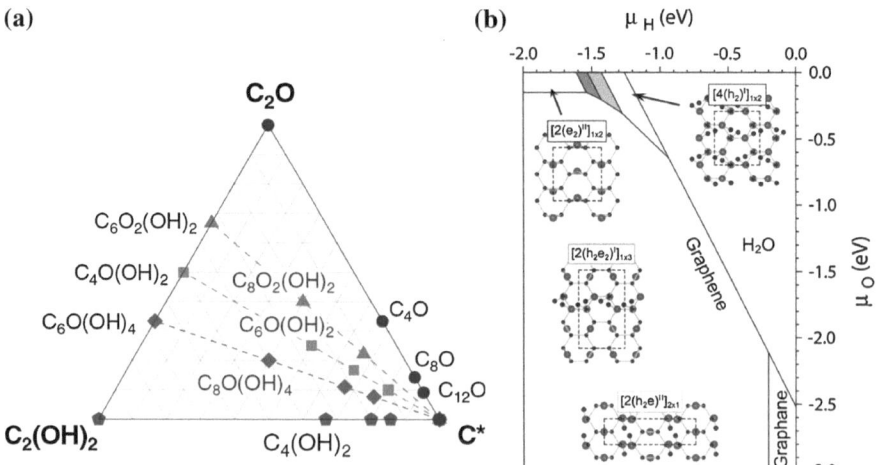

Fig. 3.4 a Ternary diagram showing ordered phases on the graphene surface with different
amounts of sp^2 Carbon (C*), epoxide (C_2O), and the 1,2-hydroxyl pair [$C_2(OH)_2$]. Reprinted
with permission from Ref. [22]. Copyright (2010) American Physical Society. **b** Thermodynamic
stability diagram of the GO phases with respect to the chemical potentials of oxygen (μ_O) and
hydrogen (μ_H). *Insets* show the atomic structures of the corresponding GO phases. Reprinted
with permission from Ref. [24]. Copyright (2010) American Physical Society

hydrogen bonds. During MC evolution, various defects are observed. For example, small holes due to break of C–C bonds are formed, where the resulting dangling bond carbons further form carbonyl and alcohol groups. In addition, other molecules such as carbon monoxide (CO), peroxide (H_2O_2) and water (H_2O), are generated. Bagri et al. [29] investigated the structural evolution of GO during thermal reduction through molecular dynamics (MD) simulations to uncover the interplay between carbon and oxygen and the degree of defects and translational order of the residual atoms at different temperatures. They found that thermal reduction will lead to the formation of carbonyl and ether groups, both of which are stable thermodynamically and cannot be removed without destroying the parent graphene sheet. Moreover, compared with the initial structure (constructed on the basis of Cai et al.'s experiment [16]), significant atomic rearrangement takes place and the GO sheets are substantially disordered after thermal annealing at 1,500 K, as shown in Fig. 3.5a, b. Especially, carbonyls, linear carbon chains, ether rings (such as furans, pyrans and pyrones), 1,2-quinones, 1,4-quinones, five-member carbon rings, three-member carbon rings and phenols, along with increased sheet roughness, are observed, as illustrated in Fig. 3.5c–l.

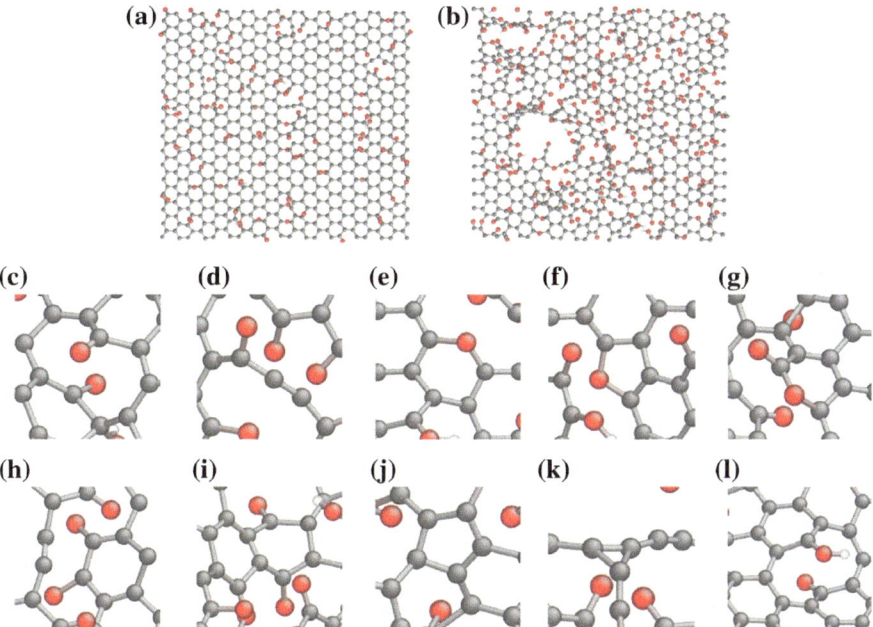

Fig. 3.5 Morphology of RGO and the structure of defects formed during thermal annealing. The RGO sheets with an initial oxygen concentration of 20 % (**a**) and 33 % (**b**) in the form of hydroxyl and epoxide groups in the ratio of 3/2 after annealing at 1,500 K. Oxygen functional groups and carbon arrangements formed after annealing: a pair of carbonyls (**c**), carbon chain (**d**), pyran (**e**), furan (**f**), pyrone (**g**), 1,2-quinone (**h**), 1,4-quinone (**i**), five-member carbon ring (**j**), three-member carbon ring (**k**) and phenol (**l**). Carbon, oxygen and hydrogen atoms are highlighted in *grey*, *red* and *white*, respectively. Reprinted with permission from Ref. [29]. Copyright (2010) Nature Publishing Group

Besides, the hydroxyl groups are easily to be desorbed at low temperatures without altering the graphene basal plane. Meanwhile, the isolated epoxy groups are relatively more stable, but substantially distort the graphene lattice on desorption. Removal of carbon from the graphene plane is more likely to occur when the initial hydroxyl and epoxide groups are in close proximity to each other. What's more, these theoretical results are corroborated by FTIR and XPS experiments.

Employing the density functional tight-binding (DFTB) method, Samarakoon and Wang [30] searched the stable structure of GO and proposed a twist-boat conformation as the energetically most favorable nonmetallic configuration for fully oxidized graphene. The fully-oxidized GO with randomly decorated hydroxyl and epoxide groups closely resembles the fully hydrogenated graphene in which there exists broad distribution of corrugated membranes that can be classified in accordance to the chair, boat, and twist-boat conformations of graphene. From the comparison of binding energies, the twist-boat along the armchair direction is the lowest-energy conformation for fully-oxidized GO. The calculated Raman G-band blue shift of this lowest-energy conformation is in good agreement with experimental observations [31, 32], which further confirms its stability.

In succession, Liu et al. [33] proposed an amorphous structural model for GO with locally ordered structural motifs by Wang et al. [24]. Based on some basic structural rules, they randomly placed epoxide and hydroxyl groups on the graphene basal plane using the MC method to build the amorphous GO structures with

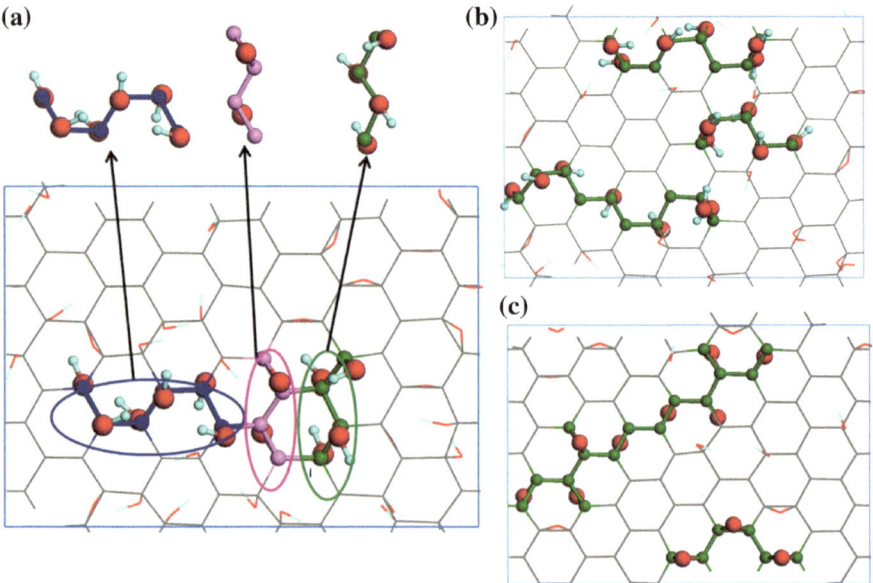

Fig. 3.6 Thermodynamic stable structures of amorphous GO: **a** GO with OH:O = 2 and $R = 70\%$; **b** GO with OH:O = 0.22 and $R = 50\%$; and **c** GO with OH:O = 8 and $R = 50\%$. The highlighted parts are the locally ordered motifs. Reprinted with permission from Ref. [33]. Copyright (2011) Elsevier Ltd.

different coverage and OH:O ratios. Then, employing the DFT calculations, these initially built structures were fully optimized. During optimization, dissociation of epoxide, formation and release of H_2O molecules, breaking of carbon-rings on the basal plane, and other structural damages were observed in some of the random GO configurations, especially for those with relatively high coverage rates ($R \geq 30$ %). After optimization, the energetically preferable structure for the amorphous GO always contains some local ordered motifs, as shown in Fig. 3.6a. In the ordered motifs, epoxide and hydroxyl groups distribute closely together, leading to formation of hydrogen bonds and lowering the total energy of the system, consistent with previous theoretical predictions [21–24]. Moreover, in these locally ordered motifs, clusters completely formed by epoxide and (or) hydroxyl groups were observed (Fig. 3.6b, c), in line with the experimental observations [18, 34, 35]. Finally, formation of GO through oxidation of graphene is an exothermic process. As the ratio of hydroxyl to epoxide groups (OH:O) increases, the GO becomes more stable.

3.2 Physical Properties

3.2.1 Thermodynamic Stabilities

Usually, GO is thermally unstable. Upon heating even below 100 °C, GO slowly decomposes possibly due to release of the absorbed water. The major mass loss occurs at ~200 °C presumably because of decompositions of oxygen-containing groups [11, 36]. However, the removal of functional groups greatly increases the thermal stability of the RGO. When the GO is heated up to 800 °C, no significant mass loss is detected [36].

Using time-resolved dynamic light scattering, Chowdhury et al. [37] investigated the aggregation kinetics and stability of GO in different aquatic chemistries (pH, salt types, and ionic strength) relevant to natural and engineered systems. It was found that GO is highly stable in both natural and synthetic surface waters. More than 90 % of GO remained suspended in the Calls Creek water (with a fairly stable hydrodynamic diameter of ~250–300 nm) after 1 month, although it settled quickly in synthetic groundwater. While the effect of pH on the stability of GO is quite complex, depending on the detailed ionic liquid [37–39].

Eigler et al. [40] investigated the effect of NaOH and HCl on the stability of carbon framework in GO after substitution or etherification reaction. They found that the carbon lattice is stable in GO treated by acid or base at 10 °C, but it is ruptured by the treatment of base at 40 °C. Moreover, after the base treatment at 10 °C, GO bears predominately hydroxyl groups; while at 40 °C in water or acidic dispersion, ether groups form. Moreover, they suggested that it is necessary to distinguish between the functional group stability and the carbon framework stability of GO [41]. As for the GO with an almost intact σ-framework of carbon atoms (ai-GO), the carbon framework can be stable up to 100 °C while functional groups have already transformed, as measured by the statistical Raman microscopy.

But for the GO prepared by the Hummers' method (GO-c), it is less stable and starts to decompose at 50 °C by releasing CO_2 gas.

In addition, employing DFT calculations, it was found that reduction of GO from high coverage to low coverage, such as from 75 to 6.25 %, is relatively easy, but further reduction is rather difficult [19]. This agrees with Stankovich et al.'s report that at certain condition, further heating up GO results in no significant mass loss [36]. Also, the thermodynamic stability of GO relies on its detailed geometry structure and chemical stoichiometry. Due to significant local distortion, a single functional group adsorbed on graphene has an adsorption energy of –4.72 eV for epoxide and –9.34 eV for hydroxyl [22]. Adsorption of the functional groups to form pairs on both sides of the graphene can further stabilize GO. Especially, it is energetically favorable for the hydroxyl and epoxide groups to aggregate together and to form specific types of chains with sp^2 carbon regions in between [21, 22]. Moreover, the reaction of graphene being oxidized to GO is an exothermic process with negative formation energy [33]. As the coverage of functional groups increases, more and more graphene will be oxidized, leading to lower formation energy. Therefore, the fully-oxidized GO is usually energetically preferable [24], as supported by the GO phase diagram shown in Fig. 3.4b. Meanwhile, increase of the OH:O ratio will lead to enhanced stability of GO since large population of hydroxyls will form more hydrogen bonds [33], as shown in Fig. 3.7a. This hydroxyl-rich GO structure agrees with Kim et al.'s excremental characterization [42] that the as-synthesized multilayer GO films are rich in epoxy groups and will evolve toward hydroxyl-rich GO at room temperature.

Besides, through MD simulations, Kumar et al. [43] also demonstrated that as-synthesized RGO commonly with large fractions of oxygen-rich epoxide and carbonyl groups is only kinetically metastable at room temperature. Once exposed to lower temperatures and oxygen partial pressures than those used during the reduction process, the metastable oxygen-rich RGO will transforms to hydroxyl-rich structure with lower oxygen content, driven by the carbonyl to hydroxyl

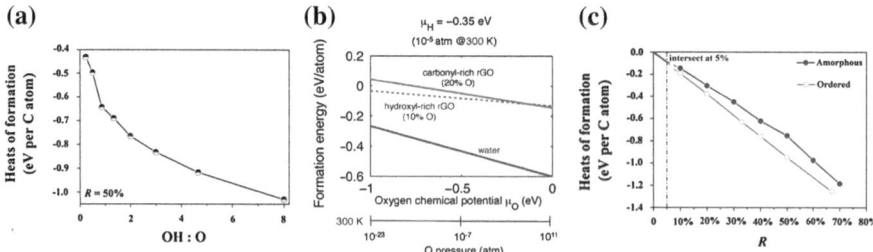

Fig. 3.7 Heats of formation for the GO with $R = 50$ % but different OH:O ratios (**a**) and for the GO with OH:O = 2 but different coverages (**c**). Reprinted with permission from Ref. [33]. Copyright (2011) Elsevier Ltd. **b** Formation energy of hydroxyl-rich (10 % O) and carbonyl-rich (20 % O) RGO structures as a function of oxygen chemical potential. The standard value of atmospheric hydrogen chemical potential at 300 K is −0.35 eV. Reprinted with permission from Ref. [43]. Copyright (2013) American Chemical Society

conversion near the carbon vacancies and holes. The mechanism involved in this spontaneous reduction of RGO can be addressed by water formation via the interaction of hydrogen and oxygen atoms within the basal plane, a process observed to be favorable at room temperature and under ambient atmospheric hydrogen and oxygen partial pressures, as displayed in Fig. 3.7b.

By comparing the heats of formation between the ordered and amorphous GO with the same chemical stoichiometry, Liu et al. [33] found that the ordered GO is more stable than the amorphous one in a certain chemical stoichiometry. For GO with OH:O ratio of 2, as the coverage of functional groups increases, the energy difference between the ordered and amorphous GO enlarges accordingly. However, at low coverage, the energy difference between the ordered and amorphous GO is negligible and the amorphous GO can be as stable as the ordered one, especially for the coverage less than 5 %, as illustrated in Fig. 3.7c.

3.2.2 Mechanical Properties

Exploring the mechanical properties of GO is of great importance since they are closely related to the reliability and service life of GO-based nanodevices. Due to diverse types and different coverage of the functional groups, the mechanical properties of GO are tunable. In 2007, Dikin et al. [44] measured the mechanical of GO papers, which are micrometer thick films obtained by slow evaporation of GO solution or by filtration method. After analyzed from 31 samples, the average modulus of GO papers was 32 GPa with the highest being 42 ± 2 GPa, and the intrinsic strength ranged from 15 to 133 MPa. Both of which are higher than those reported for flexible graphite foil and bucky paper. Generally, the reported Young's modulus and intrinsic strength values of GO papers show a wide distribution, ranging from 6 to 42 GPa and from 76 to 293 MPa, respectively [45].

Besides, the mechanical properties of GO papers can be tuned by doping or compositing. By doping a small amount (less than 1 wt%) of Mg^{2+} and Ca^{2+} ions, Park et al. [46] demonstrated significant enhancement in mechanical stiffness (10–200 %) and fracture strength (~50 %) of the graphite oxide, as shown in Table 3.1. Under tensile loading, the edge-bound metal ions can resist normal deformations between sheets that are on the same plane. Meanwhile, small mechanical perturbations can cause the edge-bound metal ions to adopt more favorable chemical interactions with the oxygen-containing functional groups of the GO sheets, which will also enhance the stiffness and strength of GO papers. In addition to alkaline earth metal ions, they suggested that mechanical stiffness and strength of GO papers can be also increased by chemically cross-linked polyallylamine (PAA) [47] (see Table 3.1). Moreover, introducing glutaraldehyde or water molecules into the gallery regions will effectively tailors the interlayer adhesions of the graphite oxide. Both the tensile modulus and strength show significant improvements for the glutaraldehyde-treated graphite oxide, but decreased mechanical properties are observed for the H_2O-treated graphite oxide [45].

Table 3.1 Mechanical properties of unmodified and modified GO Papers

Materials	E_I (GPa)	E_S (GPa)	E_E (GPa)	σ (MPa)	ε (%)
Unmodified paper	5.8 ± 1.4	16.6 ± 2.2	25.6 ± 1.1	81.9 ± 5.3	0.40 ± 0.03
As-prepared Mg-modified paper	15.2 ± 4.6	22.3 ± 3.1	24.6 ± 1.4	87.9 ± 14.2	0.40 ± 0.04
Rinsed Mg-modified paper	14.6 ± 0.3	21.8 ± 1.5	27.9 ± 1.8	80.6 ± 16.5	0.33 ± 0.08
As-prepared Ca-modified paper	9.8 ± 4.5	15.5 ± 2.4	21.5 ± 1.5	75.4 ± 21.6	0.41 ± 0.15
Rinsed Ca-modified paper	17.2 ± 3.4	23.3 ± 1.8	28.1 ± 1.2	125.8 ± 13.6	0.50 ± 0.06
PAA-modified paper	11.3 ± 3.2	25.5 ± 3.9	33.3 ± 2.7	91.9 ± 22.4	0.32 ± 0.08

E_I is the Young's modulus in the initial region where loading is started; E_S is the Young's modulus at $\sigma = 10$ MPa in the straightening region; E_E is the maximum Young's modulus in the linear region; σ is the intrinsic strength; and ε is the fracture strain. Reprinted with permission from Refs. [46, 47]. Copyright (2008) and (2009) American Chemical Society

By irradiating the GO aqueous dispersion with γ-rays in the presence of oxygen, Liu et al. [48] indicated part of the functional groups on GO can be transformed into peroxides, forming graphene peroxide (GPO). Then using GPO as a polyfunctional initiating and cross-linking center, graphene-based composite hydrogels with exceptional mechanical properties can be obtained, which exhibit high tensile strengths of 0.2–1.2 MPa and extremely large elongations of 2,000–5,300 %.

As mentioned above, the GO papers have a wide Young's modulus range of 6–42 GPa [45]. However, the Young's modulus of GO paper closely depends on the thickness. When the film thickness is reduced down to a few layers, its Young's modulus increases dramatically to about 200 GPa [49–51]. Especially, GO monolayer has a much larger Young's modulus than that of thick GO paper. An AFM measurement suggests that RGO monolayer has a mean Young's modulus of 250 GPa with a standard deviation of 150 GPa [50], as shown in Fig. 3.8a. Moreover, the effective modulus (E_{eff}) depends on the geometry of the samples, following the Eq. (3.3) below:

$$E_{eff} = E + 17Tl^2/\left(32wt^3\right), \tag{3.3}$$

where E is the Young's modulus; T is the tension in the sample; t, w, and l are the thickness, width, and length of the sample, respectively. For the single-layer RGO (having a fixed height), its effective modulus shows a linear dependence on the l^2/w, where an approximately constant value of ~4 nN for all the sheets can be inferred, as shown in Fig. 3.8b. Then, taking a van der Waals (vdW) thickness of 7 Å for the GO layer, Young's modulus of 207.6 ± 23.4 GPa and elastic constant of 145.3 ± 16.4 N/m were reported by Suk et al. using AFM measurement combined with finite element analysis [51]. They also studied the effective modulus of two-layer and three-layer GO sheets, as illustrated in Fig. 3.8c.

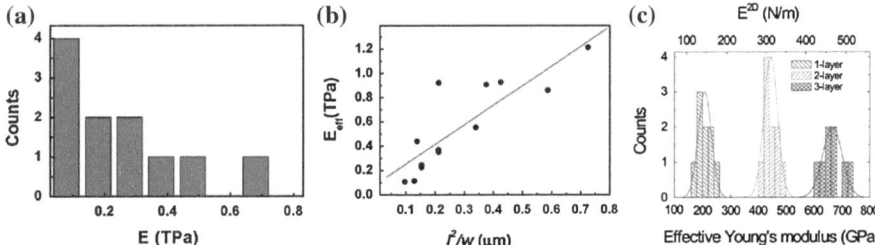

Fig. 3.8 **a** Young's modulus of the RGO; **b** relationship between the E_{eff} and l^2/w for different RGO monolayer samples. Reprinted with permission from Ref. [50]. Copyright (2008) American Chemical Society. **c** Histogram of the effective Young's modulus and two-dimensional elastic constant (E^{2D}) of GO sheets with 1-, 2-, 3-layer thickness. The *solid lines* represent Gaussian fits to the data. Reprinted with permission from Ref. [51]. Copyright (2010) American Chemical Society

Other factors such as coverage, arrangement, and ratio of the functional groups also affect the mechanical property of GO. Zheng et al. [52] studied the mechanical properties of graphene with different functional groups, such as –OH and –COOH and pointed out that Young's modulus of the functionalized graphene reduces dramatically with increasing coverage of surface functional groups, as shown in Fig. 3.9a. Meanwhile, the type of functional group will influence the Young's modulus, also shown in Fig. 3.9a. However, the molecular weight of the functional groups plays a minor role in determining Young's modulus, as illustrated in Fig. 3.9b.

Using first-principles calculations and the previously proposed structural models [33], Liu et al. [53] systematically investigated the effects of coverage, arrangement (ordered or amorphous) and OH:O ratio of the functional groups on the mechanical property of GO. The Young's modulus and intrinsic strength of GO mainly depend on the coverage and arrangement of the epoxide and hydroxyl groups, both of which decrease with increasing the coverage, as shown in Fig. 3.9c. This is consistent of Zheng's reports [52]. Meanwhile, GO systems with orderly arranged functional groups has larger Young's modulus and intrinsic strength than those of the randomly arranged ones. At certain coverage, the Young's modulus of GO only slightly fluctuates with the OH:O ratio, as shown in Fig. 3.9d. This is because once the coverage is fixed, change of OH:O ratio only slightly varies the molecular weight of the whole functional groups. As reported by Zheng et al. [52], the molecular weight of the functional groups plays a minor role in determining Young's modulus. Therefore, the theoretical finding coincides with the experimental observation. Indeed, the main factor contributing to this result is that change of OH:O will lead to variation of thickness of GO. Table 3.2 gives the Young's modulus and intrinsic strength of GO with different coverages and arrangements of the functional groups.

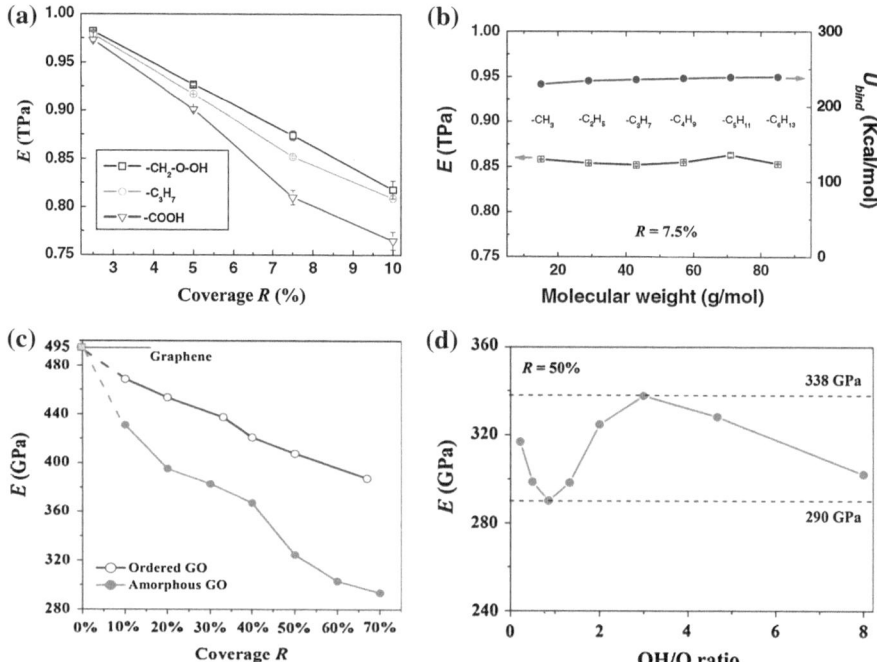

Fig. 3.9 **a** Young's modulus as a function of R for $-CH_2-O-OH$, $-C_3H_7$, and $-COOH$ functionalized graphene sheet; **b** effect of the molecular weight of functional groups on Young's modulus of graphene sheet ($R = 7.5$ %) and the corresponding binding energy between the functional groups and graphene sheet. Reprinted with permission from Ref. [52]. Copyright (2010) Elsevier Ltd. **c** Young's modulus as a function of R for both ordered and amorphous GO structures. The scaled values for perfect graphene sheet by assuming a vdW distance of 7 Å (*filled square*) are given for reference. **d** Relationship between Young's modulus and OH/O ratio for the amorphous GO structures with $R = 50$ %. The *dashed lines* bracket the range of Young's modulus. Reprinted with permission from Ref. [53]. Copyright (2012) The Royal Society of Chemistry

Table 3.2 The Young's modulus (E) and intrinsic strength (σ) of GO with OH:O $= 2$ but different coverage (R) and arrangements (ordered or amorphous) of the functional groups

R (%)	E (GPa)		σ (GPa)	
	Ordered GO	Amorphous GO	Ordered GO	Amorphous GO
0	495.0	495.0	47.8	47.8
10	468.6	430.9	46.3	40.9
20	453.6	395.3	44.4	37.5
40	420.9	367.4	40.0	33.1
50	407.7	324.7	38.6	27.9

Notice that the E and σ for the case of $R = 0$ % are the scaled values of graphene with a vdW distance of 7 Å. Reprinted with permission from Ref. [53]. Copyright (2012) The Royal Society of Chemistry

3.2.3 Electronic Properties

The experimentally fabricated GO samples are usually insulating. Typically, GO has a sheet resistance of ~10^{10} Ω sq^{-1} or higher due to large population of sp^3 hybridized carbons bonded with the oxygen-containing groups [54–57]. However, the sheet resistance can be lowered by reduction of GO. Gilje et al. [54] reported that the measured sheet resistance of the GO film of ~4 × 10^{10} Ω sq^{-1} and the RGO value of ~4 × 10^6 Ω sq^{-1}. Using the vacuum filtration method, Eda et al. [55, 56] was able to achieve uniform thin films with a controllable number of GO layers over large areas. For the hydrazine-treated GO thin films, the sheet resistance is nearly independent of the filtration volume except at very high values (>300 ml). However, after annealing at 200 °C in nitrogen (or vacuum), the sheet resistance dramatically decreases with the filtration volume reduces from ~10^{11} to ~10^5 Ω sq^{-1}, as shown in Fig. 3.10a. Becerril et al. [57] characterized the sheet resistance and optical transparency of GO thin films obtained using different reduction treatments. Compared with the non-reduced GO, both chemical

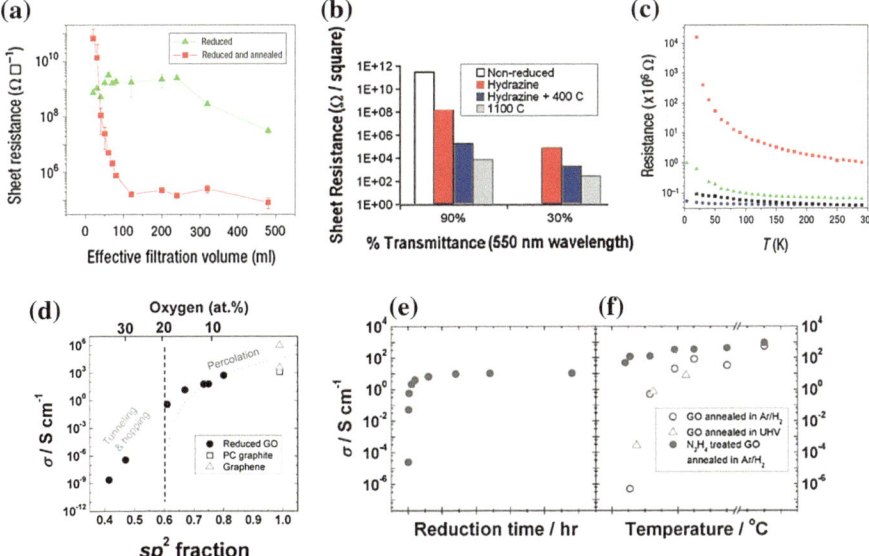

Fig. 3.10 a Sheet resistance at λ = 550 nm as a function of filtration volume for RGO films. Reprinted with permission from Ref. [55]. Copyright (2008) Nature Publishing Group. **b** Sheet resistance of GO undergoing different reduction treatments. Reprinted with permission from Ref. [57]. Copyright (2008) American Chemical Society. **c** Resistance as a function of temperature for the as-made graphene (*green*), graphene annealed at 800 °C with a titanium/gold contact (*black*) and with a palladium contact (*blue*), and GO annealed at 800 °C (*red*). Reprinted with permission from Ref. [58]. Copyright (2008) Nature Publishing Group. Electric conductivity of GO as a function of residual oxygen and sp^2 fraction (**d**), reduction time (**e**), and temperature (**f**). Reprinted with permission from Ref. [59]. Copyright (2009) WILEY-VCH Verlag GmbH & Co. KGaA, Weinheim

and thermal reduction treatments will decrease the sheet resistance, as shown in Fig. 3.10b, where a thermal graphitization procedure was most effective, producing films with sheet resistances as low as 10^2–10^3 Ω sq^{-1} with 80 % transmittance for 550 nm light. In addition, the resistance of RGO sheets is closely related to the annealing temperature [58], as shown in Fig. 3.10c.

As inverse of the resistance, the electric conductivity is an important parameter to characterize the electronic properties of GO sheets. As shown by Becerril et al. [57], the conductivity of GO film increases during reduction. Mattevi et al. [59] studied the role of residual oxygen and the sp^2 bonding fraction on the conductivity of GO. As the GO is gradually reduced, the oxygen concentration decreases due to removal of functional groups, but the fraction of sp^2 carbon increases, which further leads to an enhanced conductivity, as shown in Fig. 3.10d. Moreover, the electric conductivity also depends on reduction time and temperature [59]. As the reduction time and temperature increases, conductivity of GO increases accordingly and finally reaches a saturated value, as shown in Fig. 3.10e, f.

The electronic band gap of GO sheets can be also tuned by change of oxidation degree [60]. Guo et al. [61] demonstrated that through reduced by femtosecond laser pulses, oxygen contents of GO in the reduced region could be modulated by varying the laser power. In fact, the band gap of reduced GO was precisely modulated from 2.4 to 0.9 eV by tuning the femtosecond laser power from 0 to 23 mW, as shown in Fig. 3.11a. On the other hand, Ito et al. [62] studied the GO with only epoxide groups using first-principles calculations. They found that as O/C ratio increase from 0 to 50 %, the band gap increases significantly from 0 to 3.39 eV. Then, Nourbakhsh et al. [63] further confirmed such trend of band gap increase with O/C ratio. For perfect graphene (C_{18}), the calculated band structure shows a linear band intersecting at Fermi energy without band gap. However, after adsorbing the oxygen ($C_{18}O_2$ configuration), a band gap of 0.2 eV appears. Further increase in the oxygen density (27.8 % with a configuration $C_{18}O_5$) results in a band gap of 1.4 eV. For an oxygen density as high as 50 % ($C_{18}O_9$), the band gap raises to 3.6 eV. Both Ito et al. [62] and Nourbakhsh et al. [63] show a monotonic increase of the band gap with the oxygen density. However, a detailed study carried out by Huang et al. [64] suggested that the band gap of doesn't increase monotonically with the oxygen density, as shown in Fig. 3.11b. In addition, Xiang et al. [27] pointed out that for the GO with only epoxide groups, different edge structures exhibit different electronic structures. For GO with an armchair edge between bare graphene and the oxidized phase, there is a band gap opening, as shown in Fig. 3.11c. In contrast, for GO with a zigzag edge between bare graphene and the oxidized phase, it is metallic, as presented in Fig. 3.11d.

Theoretically, band gap of GO with both the epoxide and hydroxyl groups has also been considered. The same tendency of band gap variation with the proportion of oxygen-containing groups was found. Analysis of electron density of states (DOS) for GO shows that the energy gap decreases from 2.8 to 1.8 eV as the coverage of functional groups drops from 75 to 50 % [19]. Employing the LDA calculation, Yan et al. [21, 22] indicated that band gap of the GO can be tuned in

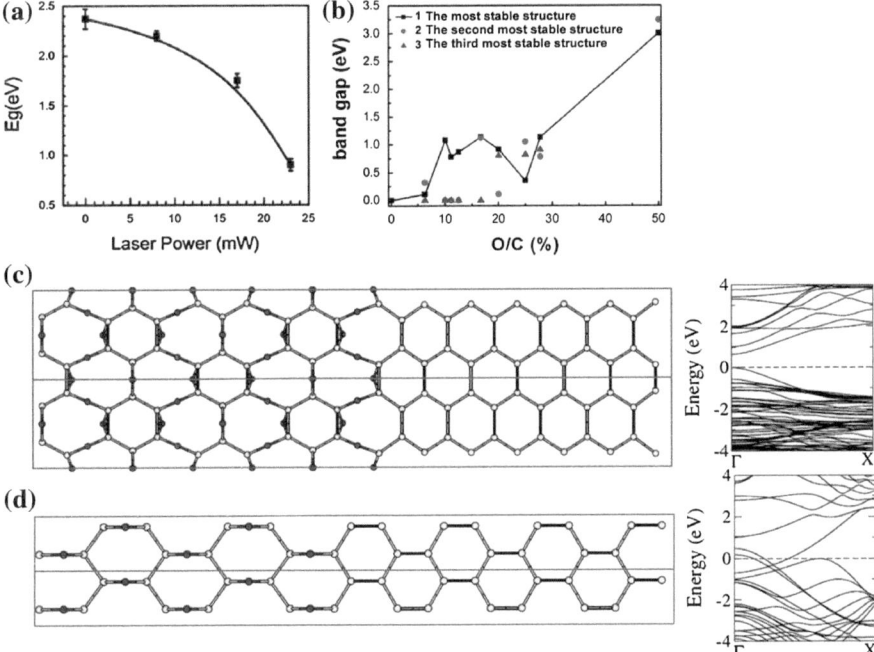

Fig. 3.11 a Dependence of band gap on the reduction laser power. Reprinted with permission from Ref. [61]. Copyright (2012) American Chemical Society. **b** Band gap of the RGO versus O/C ratio. Reprinted with permission from Ref. [64]. Copyright (2012) AIP Publishing LLC. **c** Structure of GO with an armchair edge between bare graphene and the oxidized phase and its band structure; **d** structure of GO with a zigzag edge between bare graphene and the oxidized phase and its band structure. Reprinted with permission from Ref. [27]. Copyright (2010) American Physical Society

a large range of 0–4.0 eV by varying the coverage of oxygen-containing groups. This result is also supported by Liu et al.'s calculations [33]. For both the ordered and amorphous GO structures, the band gap increases with the coverage of functional groups. As shown in Fig. 3.12a, the band gaps between the conduction band minimum (CBM) and valence band maximum (VBM) of the ordered GO structures with $R = 20$, 40, and 50 % are 1.22, 1.92 and 2.06 eV, respectively, showing a clear increase with coverage. Although there are some defect states near the Fermi energy for the amorphous GO structures, their band gaps also show an overall rising trend with the coverage, as shown in Fig. 3.12b. Since the defect-induced states are almost localized and hardly contribute to the electric conductivity, band gaps of amorphous GO structures with $R = 20$, 40, and 50 % could be counted as 0.53, 0.80 and 1.77 eV, respectively by ignoring the defect-induced states. In addition, the electronic property of GO can be affected by the OH:O ratio. As the OH:O ratio increases, the band gap increases accordingly due to enhanced degree of sp^3 hybridization [33].

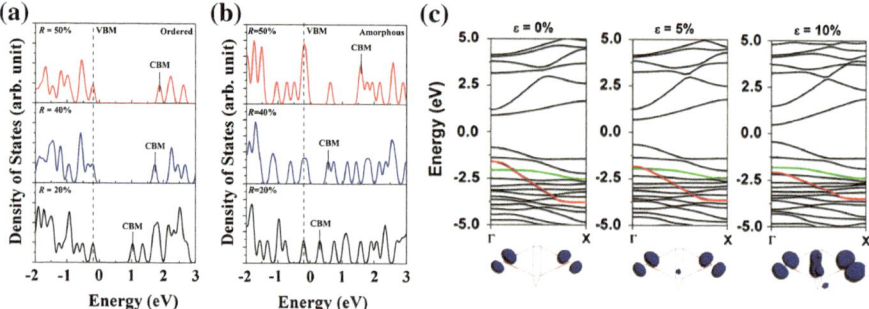

Fig. 3.12 DOS of the ordered (**a**) and amorphous (**b**) GO as a function of coverage R. Reprinted with permission from Ref. [33]. Copyright (2011) Elsevier Ltd. **b** Band structures and orbital variation of the ordered GO under tensile strain with OH/O = 2.00 and R = 50 %. Reprinted with permission from Ref. [53]. Copyright (2012) The Royal Society of Chemistry

Particularly, GO exhibits significant electromechanical effect [53]. As the GO is uniaxially elongated, C–O hybridization becomes weaker and more electrons are released, resulting in a reduction of band gap, as shown in Fig. 3.12c. For an ordered GO with coverage of 50 % and OH:O = 2.00, the band gap shrinks from 1.41 to 0.61 eV with a reduction extent of ~57 % under a tensile strain from 0 to 10 %. For an amorphous GO with the same stoichiometry, when it is undergoing a tensile strain from 0 to 8 %, the band gap shrinks from 1.03 to 0.78 eV with a reduction extent of ~24 %.

3.2.4 Optical Properties

Generally, GO monolayer is transparent with high optical transmittance in the visible spectrum region due to the atomically thin nature [65]. The GO films made from 0.5 mg/ml suspensions have an optical transmittance value of 96 % at a wavelength of 550 nm [66]. Moreover, the optical transmittance of GO films can be continuously tuned by varying the film thickness or the extent of reduction. It was reported that reduced thin GO films (with a thickness less than 30 nm) is semitransparent [67], while the atomically thin GO can be highly transparent [65, 66]. Becerril et al. [57] studied the effect of thickness on the transmittance in detail. As shown in Fig. 3.13a, when the thickness of GO films decreases from 41 to 6 nm, the optical transmittance increases clearly from ~20 to ~90 %. Besides, reduction of GO will decrease the optical transmittance [55, 57, 59]. The non-reduced GO of 9 nm thickness has a transmittance of ~98 %, but RGO of 6 nm thickness has a lower value of ~90 % (see Fig. 3.13a). Meanwhile, as the sheet resistance of RGO increases, its transmittance at λ = 550 nm rises dramatically at the early stage (with the sheet resistance less than $10^7\ \Omega\ \mathrm{sq}^{-1}$) and then gradually reach a maximum value when the sheet resistance further increases [56].

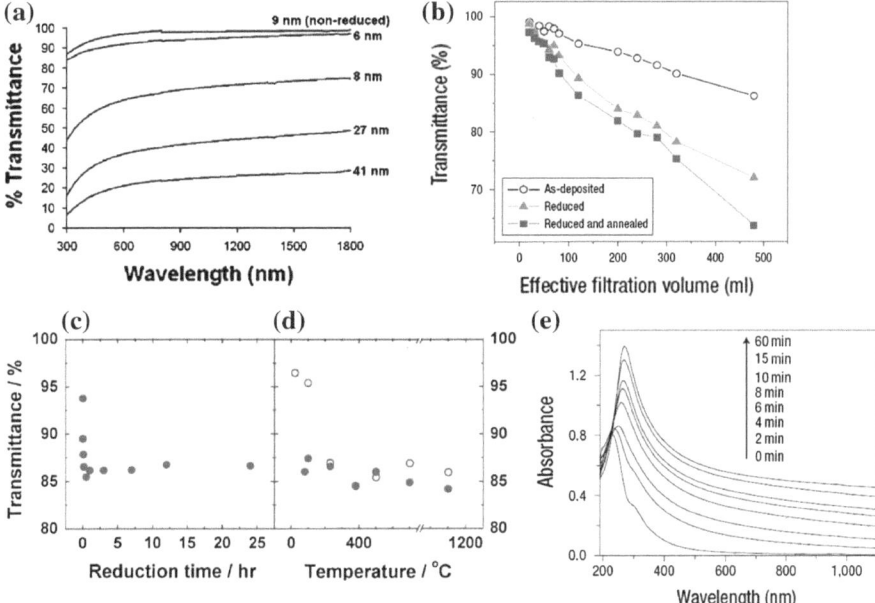

Fig. 3.13 **a** Optical transmittance spectra of the GO films in with different film thickness. Reprinted with permission from Ref. [57]. Copyright (2008) American Chemical Society. **b** Transmittance at $\lambda = 550$ nm as a function of filtration volume for RGO films with different reduction steps. Reprinted with permission from Ref. [55]. Copyright (2008) Nature Publishing Group. Transmittance at $\lambda = 550$ nm as functions of reduction time (**b**) and temperature (**c**). Reprinted with permission from Ref. [59]. Copyright (2009) WILEY-VCH Verlag GmbH & Co. KGaA, Weinheim. **e** UV-vis absorption spectra showing the change of the absorption peak of GO dispersions as a function of reaction time. Reprinted with permission from Ref. [68]. Copyright (2008) Nature Publishing Group

As shown in Fig. 3.13b, the corresponding transmittances as a function of the filtration volume at $\lambda = 550$ nm for the as-deposited GO, chemically reduced GO and chemically reduced and annealed GO are plotted [55]. It can be seen that the chemically reduced and annealed GO leads to a decrease in the transparency of thin films that is lower than that for reduced and non-annealed GO. Similarly, as shown in Fig. 3.13c, d, a long-time chemical reduction and a high-temperature thermal annealing of GO cause a progressive decrease in transparency of the thin films, reaching a saturation value of ~85 %.

The effect of reduction on the transmittance can be ascribed to the increased concentration of π electrons. This is supported by the UV–vis–IR spectroscopy study carried out by Li et al. [68]. From Fig. 1.13e it can be noticed that as the reduction time increases, absorption peak of the GO dispersion at 231 nm gradually red shifts to 270 nm and the absorption region increases with reaction time, suggesting that the electronic conjugation within the graphene sheets is restored upon hydrazine reduction. Since reduction of GO will increase the proportion of sp^2 carbon due to removal of functional groups, it can be inferred that the optical absorption of GO is dominated by $\pi-\pi^*$ transitions, which typically give rise to an absorption peak between 225 and 275 nm (4.5–5.5 eV) [67]. The absorption peak near 230 nm is featured by the $\pi-\pi^*$ plasmon and a shoulder around 300 nm is often attributed to $n-\pi^*$ transitions of C=O.

Employing DFT calculations, Johari and Shenoy [69] investigated the effects of different functional groups on the optical properties of GO structures. As presented in Table 3.3, a series of GO structures with different coverage and functional groups (epoxy, hydroxyl, and carbonyl) are constructed and studied. As shown in Fig. 3.14, according to the calculated electron energy loss spectra (EELS), the concentration of epoxy and hydroxyl functional groups varies from 25 to 75 %, and the $\pi + \sigma$ plasmon peak shows a clear blue shift of about 1.0–3.0 eV. The π plasmon peak is less sensitive to the concentration of epoxy and hydroxyl functional groups and exhibits a blue shift of ~0.4 eV. Different from the effect of epoxy and hydroxyl functional groups, increase of the carbonyl groups in the center of graphene sheet will create holes, which lead to a red shift of the EELS. In the case of 37.5 % of oxygen-to-carbon ratio, the π plasmon peak is about 4.0 eV, which is red shift of 0.8 eV compared to that of the pristine graphene (4.8 eV). Then, through a controllable stepwise reduction of GO at room-temperature, Mathkar et al. [70] suggested that during reduction, the optical gap of GO can be tuned from 3.5 eV down to 1.0 eV with a concurrent increase of C/O ratio. Meanwhile, they are also able to identify reduction of functional groups by analyzing signature IR absorption frequencies. Carbonyl group is the first to be reduced, while the tertiary alcohol takes the longest time to be completely removed from the GO surface.

Another important optical property of GO is its intrinsic fluorescence in the near-infrared (NIR), visible and ultraviolet regions [65, 71–74], which is the most notable difference from graphene since there is no fluorescence in graphene due to absence of an energy gap [75]. As shown in Fig. 3.15a, fluorescence measurements in the visible range of the nano-GO (NGO) with a dimensional size down to 10 nm demonstrates that the NGO has an emission peak at ~570 nm at 400 nm excitation [76]. When the NGO was bonded with 6-arm branched PEG molecules (NGO-PEG), the emission peak was blue-shifted to ~520 nm, which might be ascribed to the nanosize and change of the functional groups of the NGO under reduction. Besides, photoluminescence of both NGO and NGO-PEG in the IR and NIR regions was also discovered, as shown in Figs. 3.15b, c. Similarly, Luo et al. [72] also reported broadband visible photoluminescence of GO, which can be further modified by progressive chemical reduction, as shown in Fig. 3.15d. The broad photoluminescence indicates a dispersion of gaps, which may arise from bond alternation within the GO plane giving rise to intervalley scattering. In this case, a "Kekule pattern" emerges in the electronic potential, providing a spatially modulated intervalley gap parameter.

Furthermore, weak blue to ultraviolet fluorescence was observed in the as-synthesized GO thin films (centered around 390 nm) and solutions (centered around 440 nm) under ultraviolet radiation [73, 77, 78], as shown in Fig. 3.15e. By appropriately controlling the concentration of isolated sp^2 clusters through

Table 3.3   A series of GO structures with different coverages and functional groups, itemed from a to n (Gr stands for pristine graphene)

Structure	No. of atoms in the unit cell	sp^3 bonded carbon atoms (%)	Functional groups			k point mesh		E_{form} (eV)
			Epoxy	Hydroxyl	Carbonyl	vac $\approx$ 12 Å	vac $\approx$ 27 Å	
Gr	4	0	0	0	0	45 × 45 × 1	45 × 45 × 1	–
a	9	25	1	0	0	35 × 25 × 1	25 × 15 × 1	0.13
b	10	50	2	0	0	35 × 25 × 1	25 × 15 × 1	0.08
c	10	50	2	0	0	35 × 25 × 1	25 × 15 × 1	–0.16
d	11	75	3	0	0	35 × 25 × 1	25 × 15 × 1	–1.09
e	11	75	3	0	0	35 × 25 × 1	25 × 15 × 1	–0.82
f	6	100	2	0	0	45 × 45 × 1	20 × 40 × 1	–0.94
g	6	100	2	0	0	45 × 45 × 1	20 × 40 × 1	–0.76
h	13	50	1	2	0	35 × 25 × 1	25 × 15 × 1	–3.53
i	14	75	2	2	0	35 × 25 × 1	25 × 15 × 1	–4.67
j	42	50	2	8	0	8 × 32 × 1	8 × 32 × 1	–17.04
k	24	100	4	4	0	21 × 42 × 1	10 × 20 × 1	–11.17
l	10	0	0	0	2	35 × 25 × 1	25 × 15 × 1	–0.47
m	16	0	0	0	4	30 × 15 × 1	20 × 10 × 1	–2.24
n	22	0	0	0	6	30 × 15 × 1	12 × 8 × 1	–4.09

Formation energy (E_{form}) per unit cell and k-points mesh used for the relaxation and the electronic structure calculations (vac $\approx$ 12 Å) and for the loss spectra calculations (vac $\approx$ 27 Å) were also given. Reprinted with permission from Ref. [69]. Copyright (2011) American Chemical Society

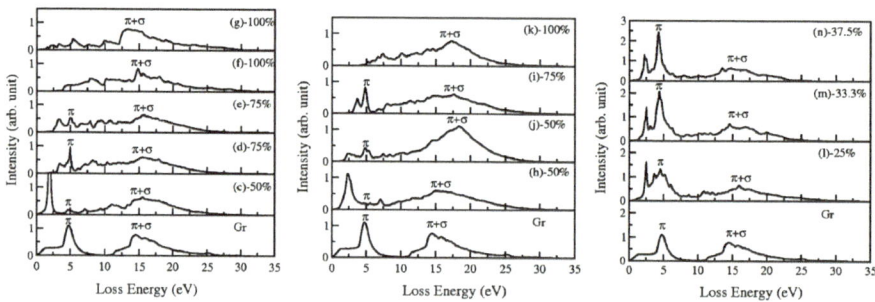

Fig. 3.14 EELS of GO with epoxy (*left*), epoxy and hydroxyl (*center*), and carbonyl (*right*) functional groups. The label number of each structure corresponds to the one listed in Table 3.3. Reprinted with permission from Ref. [69]. Copyright (2011) American Chemical Society

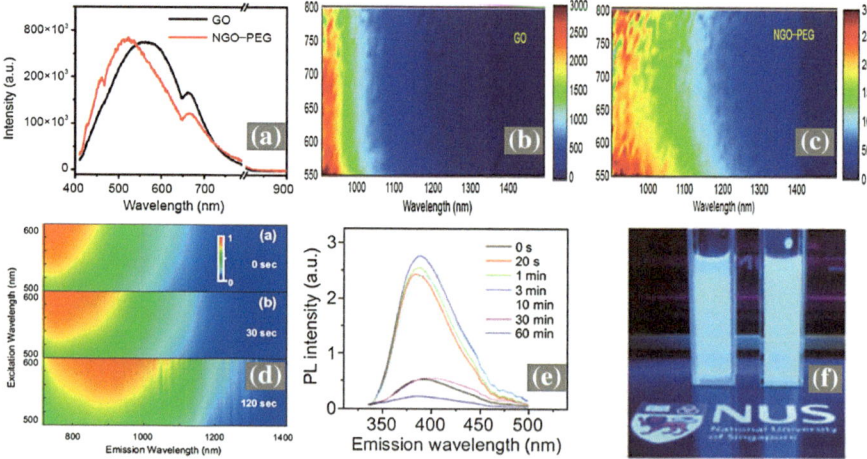

Fig. 3.15 **a** Fluorescence of GO (*black curve*) and NGO-PEG (*red curve*) in the visible range under an excitation of 400 nm. Photoluminescence excitation spectra of GO (**b**) and NGO-PEG (**c**) with 0.31 mg/mL graphitic carbon in the IR region. Reprinted with permission from Ref. [76]. Copyright (2008) Tsinghua Press and Springer-Verlag GmbH. **d** Normalized photo-luminescence excitation-emission maps for solid GO taken in transmission during hydrazine vapor exposure of different time. Reprinted with permission from Ref. [72]. Copyright (2009) AIP Publishing LLC. **e** Fluorescence spectra of a GO thin-film excited at 325 nm after exposure to hydrazine vapour for different periods of time. Reprinted with permission from Ref. [73]. Copyright (2010) WILEY-VCH Verlag GmbH & Co. KGaA, Weinheim. **f** *Blue* fluorescence for supernatant solution of nanosized GO. Reprinted with permission from Ref. [79]. Copyright (2009) American Chemical Society

reduction treatment, the photoluminescence intensity can be increased by a factor of ten compared to the as-synthesized GO films [73]. Similar blue fluorescence was also observed in water-soluble GO fragments produced by ionic-liquid-assisted electrochemical exfoliation of graphite [79], as shown in Fig. 3.15f. Both

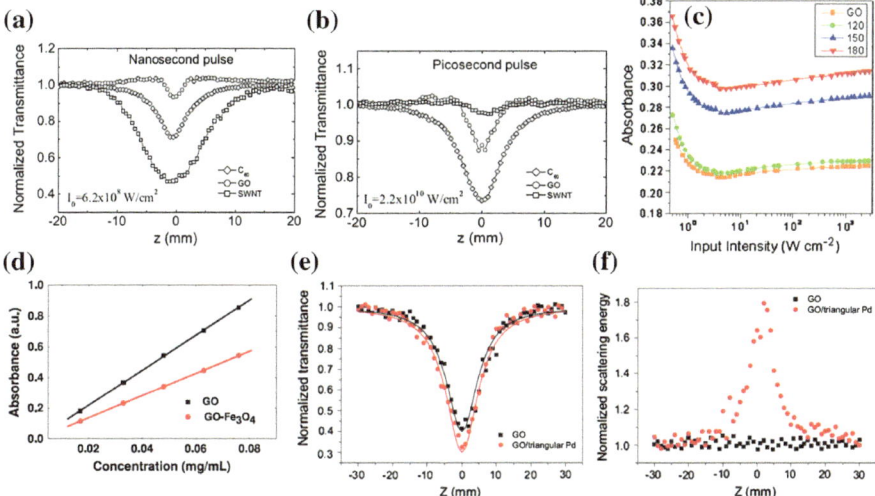

Fig. 3.16 Open aperture Z-scan curves of GO, SWCNT in DMF (*N,N*-Dimethylformamide), and C$_{60}$ in toluene for nanosecond pulses (**a**) and picosecond pulses (**b**). Reprinted with permission from Ref. [80]. Copyright (2009) AIP Publishing LLC. **c** Nonlinear absorbance of GO where *red line* is for pristine GO, and *green*, *blue*, and *red lines* are for GO hydrothermally treated at 120, 150, and 180 °C, respectively. Reprinted with permission from Ref. [81]. Copyright (2009) American Chemical Society. **d** The plot of absorption value at 400 nm versus concentration for GO and GO-Fe$_3$O$_4$ in water. *Solid lines* are linear fits. Reprinted with permission from Ref. [83]. Copyright (2011) IOP Publishing Ltd. **e** Open-aperture Z-scans for GO and GO/triangular Pd composites where the *dots* are experimental results, and the solid lines are fitted curves. **f** Typical nonlinear scattering results for GO and GO/triangular Pd composites at 532 nm. The energy detector was located 45° from the axis. Reprinted with permission from Ref. [84]. Copyright (2014) Elsevier B.V.

the disordered geometries and the electronic structures of GO indicate that fluorescence in GO arises from recombination of electron-hole pairs in localized the electronic states originating from various possible configurations, rather than from band-edge transitions as in the case of typical semiconductors.

In addition, GO shows interesting nonlinear optical (NLO) properties [80–84]. In 2009, Liu et al. [80] studied the NLO properties of GO at 532 nm using nanosecond and picosecond pulses. As measured by the Z-scan technique, two-photon absorption dominates nonlinear absorption process of GO in the case of picosecond pulse, while GO exhibits excited state adsorption in the case of nanosecond pulse. Besides, the NLO properties of GO are distinctly different from single-wall carbon nanotube (SWCNT) and fullerene, especially in the case of picosecond pulse, as shown in Fig. 3.16a, b. Limiting action of the SWCNT is strongest in the nanosecond time scale but is poorest in picosecond regime because strong nonlinear scattering occurs in the nanosecond regime. C$_{60}$ shows strong excited state adsorption for both the nanosecond and picosecond pulses since it has large singlet and triplet excited states cross sections.

Zhou et al. [81] indicated that the NOL behavior of RGO depends on the hydrothermal treatment temperature as well as whether pulsed or continuous laser is used. As shown in Fig. 3.16c, as the hydrothermal treatment temperature increases, the absorbance increases significantly, which demonstrates that the hydrothermal treatment can create significant variation in electronic and optical properties of GO.

Similar to the mechanical properties, the NLO properties of GO can be also enhanced by compositing. For example, Zhu et al. [82] suggested that at the same level of linear extinction coefficient, GO functionalized with zinc phthalocyanine (PcZn) exhibits much larger NOL extinction coefficients and broadband optical limiting performance than GO at both 532 and 1,064 nm, indicating a remarkable accumulation effect as a result of the covalent link between GO and PcZn. Beside, Fe_3O_4 decorated GO also shows enhanced NLO properties with higher absorption value at 400 nm than that of the pristine GO in the same concentration in water [83], as presented in Fig. 3.16d. Through compositing GO with triangular Pd nanocrystals, Zheng et al. [84] reported that the optical limiting was significantly enhanced in GO/triangular Pd composites compared with the individual counterparts, as investigated using the open aperture Z-scan technique at 532 nm. The enhance NOL properties can be attributed to the addition nonlinear scattering effects due to compositing.

References

1. Hofmann, U., Holst, R.: Ber. Dtsch. Chem. Ges. B **72**, 754–771 (1939)
2. Ruess, G.: Monatsh. Chem. **76**, 381–417 (1947)
3. Scholz, W., Boehm, H.P.Z.: Anorg. Allg. Chem. **369**, 327–340 (1969)
4. He, H., Riedl, T., Lerf, A., Klinowski, J.: J. Phys. Chem. **100**, 19954–19958 (1996)
5. Lerf, A., He, H., Riedl, T., Forster, M., Klinowski, J.: Solid State Ionics **101–103**, 857–862 (1997)
6. Lerf, A., He, H., Forster, M., Klinowski, J.: J. Phys. Chem. B **102**, 4477–4482 (1998)
7. Nakajima, T., Mabuchi, A., Hagiwara, R.: Carbon **26**, 357–361 (1988)
8. Nakajima, T., Matsuo, Y.: Carbon **32**, 469–475 (1994)
9. Szabó, T., Berkesi, O., Forgó, P., Josepovits, K., Sanakis, Y., Petridis, D., Dékány, I.: Chem. Mater. **18**, 2740–2749 (2006)
10. Dreyer, D.R., Park, S., Bielawski, C.W., Ruoff, R.S.: Chem. Soc. Rev. **39**, 228–240 (2010)
11. Gao, X., Jiang, D.E., Zhao, Y., Nagase, S., Zhang, S., Chen, Z.: J. Comput. Theor. Nanosci. **8**, 2406–2422 (2011)
12. Mao, S., Pu, H., Chen, J.: RSC Adv. **2**, 2643–2662 (2012)
13. Kuila, T., Mishra, A.K., Khanra, P., Kim, N.H., Lee, J.H.: Nanoscale **5**, 52–71 (2013)
14. Chua, C.K., Pumera, M.: Chem. Soc. Rev. **43**, 291–312 (2014)
15. Lu, N., Li, Z.: Graphene oxide: theoretical perspectives. In: Zeng, J., Zhang, R.Q., Treutlein, H.R. (eds.) Quantum Simulations of Materials and Biological Systems, pp. 69–84. Springer, Dordrecht (2012)
16. Cai, W.W., Piner, R.D., Stadermann, F.J., Park, S., Shaibat, M.A., Ishii, Y., Yang, D.X., Velamakanni, A., An, S.J., Stoller, M., An, J.H., Chen, D.M., Ruoff, R.S.: Science **321**, 1815–1817 (2008)
17. Gao, W., Alemany, L.B., Ci, L.J., Ajayan, P.M.: Nat. Chem. **1**, 403–408 (2009)
18. Schniepp, H.C., Li, J.L., McAllister, M.J., Sai, H., Herrera-Alonso, M., Adamson, D.H., Prud'homme, R.K., Car, R., Saville, D.A., Aksay, I.A.: J. Phys. Chem. B **110**, 8535–8539 (2006)
19. Boukhvalov, D.W., Katsnelson, M.I.: J. Am. Chem. Soc. **130**, 10697–10701 (2008)

20. Lahaye, R., Jeong, H.K., Park, C.Y., Lee, Y.H.: Phys. Rev. B **79**, 125435 (2009)
21. Yan, J.A., Xian, L.D., Chou, M.Y.: Phys. Rev. Lett. **103**, 086802 (2009)
22. Yan, J.A., Chou, M.Y.: Phys. Rev. B **82**, 125403 (2010)
23. Wang, L., Lee, K., Sun, Y.Y., Lucking, M., Chen, Z.F., Zhao, J.J., Zhang, S.B.: ACS Nano **3**, 2995–3000 (2009)
24. Wang, L., Sun, Y.Y., Lee, K., West, D., Chen, Z.F., Zhao, J.J., Zhang, S.B.: Phys. Rev. B **82**, 161406 (2010)
25. Zhang, W.H., Carravetta, V., Li, Z.Y., Luo, Y., Yang, J.L.: J. Chem. Phys. **131**, 244505 (2009)
26. Lu, N., Yin, D., Li, Z.Y., Yang, J.L.: J. Phys. Chem. C **115**, 11991–11995 (2011)
27. Xiang, H.J., Wei, S.H., Gong, X.G.: Phys. Rev. B **82**, 035416 (2010)
28. Paci, J.T., Belytschko, T., Schatz, G.C.: J. Phys. Chem. C **111**, 18099–18111 (2007)
29. Bagri, A., Mattevi, C., Acik, M., Chabal, Y.J., Chhowalla, M., Shenoy, V.B.: Nat. Chem. **2**, 581–587 (2010)
30. Samarakoon, D.K., Wang, X.Q.: Nanoscale **3**, 192–195 (2011)
31. Kudin, K.N., Ozbas, B., Schniepp, H.C., Prud'homme, R.K., Aksay, I.A., Car, R.: Nano Lett. **8**, 36–41 (2008)
32. Liu, L., Ryu, S., Tomasik, M.R., Stolyarova, E., Jung, N., Hybertsen, M.S., Steigerwald, M.L., Brus, L.E., Flynn, G.W.: Nano Lett. **8**, 1965–1970 (2008)
33. Liu, L., Wang, L., Gao, J., Zhao, J., Gao, X., Chen, Z.: Carbon **50**, 1690–1698 (2012)
34. Yan, L., Punckt, C., Aksay, I.A., Mertin, W., Bacher, G.: Nano Lett. **11**, 3543–3549 (2011)
35. Casabianca, L.B., Shaibat, M.A., Cai, W.W., Park, S., Piner, R., Ruoff, R.S., Ishii, Y.: J. Am. Chem. Soc. **132**, 5672–5676 (2010)
36. Stankovich, S., Dikin, D.A., Piner, R.D., Kohlhaas, K.A., Kleinhammes, A., Jia, Y., Wu, Y., Nguyen, S.T., Ruoff, R.S.: Carbon **45**, 1558–1565 (2007)
37. Chowdhury, I., Duch, M.C., Mansukhani, N.D., Hersam, M.C., Bouchard, D.: Environ. Sci. Technol. **47**, 6288–6296 (2013)
38. Zhang, J., Terracciano, A., Meng, X.: Environ. Sci. Technol. **48**, 1359 (2014)
39. Chowdhury, I., Duch, M.C., Mansukhani, N.D., Hersam, M.C., Bouchard, D.: Environ. Sci. Technol. **48**, 1360 (2014)
40. Eigler, S., Grimm, S., Hof, F., Hirsch, A.J.: Mater. Chem. A **1**, 11559–11562 (2013)
41. Eigler, S., Grimm, S., Hirsch, A.: Chem. Eur. J. **20**, 984–989 (2014)
42. Kim, S., Zhou, S., Hu, Y., Acik, M., Chabal, Y.J., Berger, C., de Heer, W., Bongiorno, A., Riedo, E.: Nat. Mater. **11**, 544–549 (2012)
43. Kumar, P.V., Bernardi, M., Grossman, J.C.: ACS Nano **7**, 1638–1645 (2013)
44. Dikin, D.A., Stankovich, S., Zimney, E.J., Piner, R.D., Dommett, G.H.B., Evmenenko, G., Nguyen, S.T., Ruoff, R.S.: Nature **448**, 457–460 (2007)
45. Gao, Y., Liu, L.Q., Zu, S.Z., Peng, K., Zhou, D., Han, B.H., Zhang, Z.: ACS Nano **5**, 2134–2141 (2011)
46. Park, S., Lee, K.-S., Bozoklu, G., Cai, W., Nguyen, S.T., Ruoff, R.S.: ACS Nano **2**, 572–578 (2008)
47. Park, S., Dikin, D.A., Nguyen, S.T., Ruoff, R.S.: J. Phys. Chem. C **113**, 15801–15804 (2009)
48. Liu, J., Chen, C., He, C., Zhao, J., Yang, X., Wang, H.: ACS Nano **6**, 8194–8202 (2012)
49. Robinson, J.T., Zalalutdinov, M., Baldwin, J.W., Snow, E.S., Wei, Z., Sheehan, P., Houston, B.H.: Nano Lett. **8**, 3441–3445 (2008)
50. Gómez-Navarro, C., Burghard, M., Kern, K.: Nano Lett. **8**, 2045–2049 (2008)
51. Suk, J.W., Piner, R.D., An, J., Ruoff, R.S.: ACS Nano **4**, 6557–6564 (2010)
52. Zheng, Q., Geng, Y., Wang, S., Li, Z., Kim, J.K.: Carbon **48**, 4315–4322 (2010)
53. Liu, L., Zhang, J., Zhao, J., Liu, F.: Nanoscale **4**, 5910–5916 (2012)
54. Gilje, S., Han, S., Wang, M., Wang, K.L., Kaner, R.B.: Nano Lett. **7**, 3394–3398 (2007)
55. Eda, G., Fanchini, G., Chhowalla, M.: Nat. Nanotechnol. **3**, 270–274 (2008)
56. Eda, G., Lin, Y.Y., Miller, S., Chen, C.W., Su, W.F., Chhowalla, M.: Appl. Phys. Lett. **92**, 233305 (2008)
57. Becerril, H.A., Mao, J., Liu, Z., Stoltenberg, R.M., Bao, Z., Chen, Y.: ACS Nano **2**, 463–470 (2008)

58. Li, X., Zhang, G., Bai, X., Sun, X., Wang, X., Wang, E., Dai, H.: Nat. Nanotechnol. **3**, 538–542 (2008)
59. Mattevi, C., Eda, G., Agnoli, S., Miller, S., Mkhoyan, K.A., Celik, O., Mastrogiovanni, D., Granozzi, G., Garfunkel, E., Chhowalla, M.: Adv. Funct. Mater. **19**, 2577–2583 (2009)
60. Acik, M., Chabal, Y.J.J.: Mater. Sci. Res. **2**, 101–112 (2013)
61. Guo, L., Shao, R.Q., Zhang, Y.L., Jiang, H.B., Li, X.B., Xie, S.Y., Xu, B.B., Chen, Q.D., Song, J.F., Sun, H.B.: J. Phys. Chem. C **116**, 3594–3599 (2012)
62. Ito, J., Nakamura, J., Natori, A.: J. Appl. Phys. **103**, 113712 (2008)
63. Nourbakhsh, A., Cantoro, M., Vosch, T., Pourtois, G., Clemente, F., van der Veen, M.H., Hofkens, J., Heyns, M.M., De Gendt, S., Sels, B.F.: Nanotechnology **21**, 435203 (2010)
64. Huang, H., Li, Z., She, J., Wang, W.: J. Appl. Phys. **111**, 054317 (2012)
65. Loh, K.P., Bao, Q.L., Eda, G., Chhowalla, M.: Nat. Chem. **2**, 1015–1024 (2010)
66. Zhu, Y., Cai, W., Piner, R.D., Velamakanni, A., Ruoff, R.S.: Appl. Phys. Lett. **95**, 103104 (2009)
67. Eda, G., Chhowalla, M.: Adv. Mater. **22**, 2392–2415 (2010)
68. Li, D., Müller, M.B., Gilje, S., Kaner, R.B., Wallace, G.G.: Nat. Nanotechnol. **3**, 101–105 (2008)
69. Johari, P., Shenoy, V.B.: ACS Nano **5**, 7640–7647 (2011)
70. Mathkar, A., Tozier, D., Cox, P., Ong, P., Galande, C., Balakrishnan, K., Arava, L.M.R., Ajayan, P.M.: J. Phys. Chem. Lett. **3**, 986–991 (2012)
71. Pan, D., Zhang, J., Li, Z., Wu, M.: Adv. Mater. **22**, 734–738 (2010)
72. Luo, Z., Vora, P.M., Mele, E.J., Johnson, A.T.C., Kikkawa, J.M.: Appl. Phys. Lett. **94**, 111909 (2009)
73. Eda, G., Lin, Y.Y., Mattevi, C., Yamaguchi, H., Chen, H.A., Chen, I., Chen, C.W., Chhowalla, M.: Adv. Mater. **22**, 505–509 (2010)
74. Cuong, T.V., Pham, V.H., Tran, Q.T., Hahn, S.H., Chung, J.S., Shin, E.W., Kim, E.J.: Mater. Lett. **64**, 399–401 (2010)
75. Essig, S., Marquardt, C.W., Vijayaraghavan, A., Ganzhorn, M., Dehm, S., Hennrich, F., Ou, F., Green, A.A., Sciascia, C., Bonaccorso, F.: Nano Lett. **10**, 1589–1594 (2010)
76. Sun, X., Liu, Z., Welsher, K., Robinson, J.T., Goodwin, A., Zaric, S., Dai, H.: Nano Res. **1**, 203–212 (2008)
77. Subrahmanyam, K.S., Kumar, P., Nag, A., Rao, C.N.R.: Solid State Commun. **150**, 1774–1777 (2010)
78. Chen, J.L., Yan, X.P.: J. Mater. Chem. **20**, 4328–4332 (2010)
79. Lu, J., Yang, J.X., Wang, J., Lim, A., Wang, S., Loh, K.P.: ACS Nano **3**, 2367–2375 (2009)
80. Liu, Z., Wang, Y., Zhang, X., Xu, Y., Chen, Y., Tian, J.: Appl. Phys. Lett. **94**, 021902 (2009)
81. Zhou, Y., Bao, Q., Tang, L.A.L., Zhong, Y., Loh, K.P.: Chem. Mater. **21**, 2950–2956 (2009)
82. Zhu, J., Li, Y., Chen, Y., Wang, J., Zhang, B., Zhang, J., Blau, W.J.: Carbon **49**, 1900–1905 (2011)
83. Zhang, X.L., Zhao, X., Liu, Z.B., Shi, S., Zhou, W.Y., Tian, J.G., Xu, Y.F., Chen, Y.S.: J. Opt. **13**, 075202 (2011)
84. Zheng, C., Chen, W., Cai, S., Xiao, X., Ye, X.: Mater. Lett. **131**, 284–287 (2014)

Chapter 4
Application of GO in Electronics and Optics

Abstract The electronic properties of GO and RGO thin films can be tuned by varying the coverage of functional groups, chemical composition, film thickness and morphology, and average flake size. By appropriately tuning the deposition and reduction parameters, the GO/RGO films can be made insulating, semiconducting, or semimetallic. The tunable and controllable electronic properties of GO films enable their promising applications in electronic devices. The transparent and conducting properties of RGO can be also utilized in transparent conductors. Moreover, the excellent mechanical properties of GO films are useful for flexible electronic materials. On the other hand, as discussed in Chap. 3, GO processes outstanding optical properties. The intrinsic fluorescence of GO in wide regions leads to applications as optical sensing and detecting. The nonlinear optical properties of GO can be utilized for optical-limiting materials and saturable absorbers. Finally, existence of functional groups also brings about excellent behavior in surface enhanced Raman scattering (SERS) of GO.

4.1 GO for Electronics

4.1.1 Transparent Conductors

Due to the atomically thin layered structure, RGO is highly transparent in the visible spectrum. Moreover, highly reduced GO can be a semimetal with electronic states similar to the disordered graphene [1, 2], which shows weak changes in electrical conductance with gate voltage of the field-effect devices (ON/OFF ratio < 10) [3]. As a result, RGO could be potential candidate for transparent conductor applications as a replacement for indium tin oxide (ITO) in devices, such as organic solar cells [4, 5], organic light-emitting diodes [6] and displays [7]. It was reported that the lowest sheet resistance of RGO at transmittance of 80 % is about 1 kΩ/square, well above that of ITO and CVD graphene [3]. Becerril et al. [8] indicated that the produced RGO films exhibit sheet resistances as low as

© The Author(s) 2015
J. Zhao et al., *Graphene Oxide: Physics and Applications*,
SpringerBriefs in Physics, DOI 10.1007/978-3-662-44829-8_4

10^2–10^3 Ω/square with 80 % transmittance for 550 nm light. However, increase of the RGO film thickness will lower the optical transmittance, as shown in Fig. 4.1a. Then, Ning et al. [9] developed a simple fast room-temperature reduction strategy to convert GO films into transparent and conductive RGO films, which exhibit sheet resistances of 6.7–17.3 kΩ/square and transparencies of 75–81 % at 550 nm. Besides, through embedding silver nanowire (AgNW) into RGO films, the obtained AgNW/RGO transparent conducting electrodes show a sheet resistance as low as 27 Ω/square with transparency of 72 % at 550 nm [10]. Moreover, Wang et al. [11] demonstrated that the conducting properties of RGO thin films can be improved by optimizing the thickness and reducing at high temperatures. For RGO prepared by 1,100 °C thermal treatment with a thickness of 10.1 ± 0.76 nm, the calculated average conductivity is about 550 S/cm. When the film thickness is enlarged to 29.9 ± 1.1 nm, the conductivity of RGO film increases to 727 S/cm. At a given film thickness of ~10 nm, increase of film conductivity was observed with raising the heating temperatures from 550 to 1,100 °C.

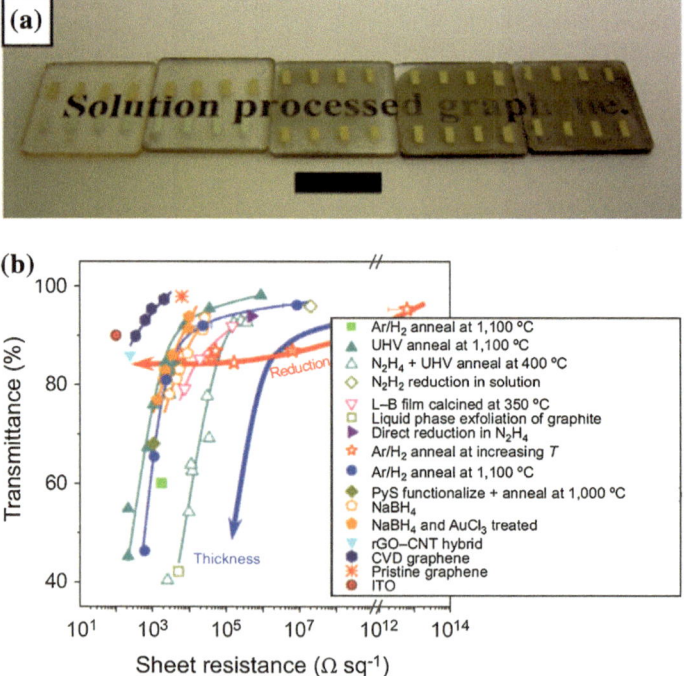

Fig. 4.1 a Photograph of an unreduced GO (*leftmost*) and a series of high-temperature reduced GO films with increased thickness. *Black scale bar* is 1 cm. Reprinted with permission from Ref. [8]. Copyright (2008) American Chemical Society. **b** Summarization of transmittance at wavelength of 550 nm as a function of sheet resistance of RGO films with different reduction and thickness reported previously. Reprinted with permission from Ref. [3]. Copyright (2010) Nature Publishing Group

So far, a range of transmittance and sheet resistance values obtained from various methods have been reported [12], including oxidation, exfoliation, dispersion, and deposition procedures, as summarized in Fig. 4.1b. With the reduction of GO and decrease of the overall film thickness, the transmittance and sheet resistance increase. Generally, efficient reduction of GO is the key to achieving highly conductive films, while optimizing the thickness of deposit films allows high optical transparencies. Currently, the highest degree of reduction is achieved via pyrolysis at 1,100 °C, yielding films with the best properties [12]. Moreover, the sheet resistance at transmittance of 90 % as low as few kΩ/square has been obtained with high temperature pyrolysis. Such sheet resistance values are approximately one order of magnitude higher than those achieved with single-walled carbon nanotube (SWCNT) thin films [13].

4.1.2 Field-Effect Devices

It is well known that graphene possesses excellent field emission properties and its field-effect mobility is one order of magnitude higher than that of Si [14]. However, the field-emission devices based on pristine graphene have relatively low I_{ON}/I_{OFF} ratios (e.g., 100 [15]) because of the finite minimum conductance of graphene at zero gate voltage. Therefore, several strategies have been developed to induce and control an electronic gap in graphene for shutting off the current, e.g., constricting its lateral dimensions to generate quasi-one-dimensional graphene nanoribbons [16–18].

Due to the unique band structure, the carriers in graphene are bipolar, that is, the electrons and holes that can be continuously tuned by a gate electrical field [15]. The bipolar feature of graphene makes its carrier type and concentration sensitive to doping. GO, the oxygen-groups functionalized graphene, therefore should improve the field emission performance of the graphene. Exceptionally low threshold field emission from atomically thin edges of RGO was reported [19]. The average threshold field required to emit currents of 1 nA was found to be less than 0.1 V/μm. Such low threshold field can be attributed to the combination of large enhancement factor and lower local work function at the edge. The edges provides an array of emission sites in the form of low work function C–O–C ether chains from which multiple electron beams are simultaneously emitted. Therefore, the enhanced field emission characteristics are attributed to the termination of straight edges by cyclic ether, which constitutes the most stable form of oxygen in RGO.

The excellent field emission properties of GO implies its promising prospect in field-effect transistors (FETs). Using chemically reduced GO (CRGO) sheets, Joung et al. [20] fabricated FETs where the CRGO sheets suspended in water are assembled between prefabricated gold source and drain electrodes using ac dielectrophoresis. When applying a backgate voltage, 60 % of the devices show p-type FET behavior, while the remaining 40 % show ambipolar behavior.

After mild thermal annealing at 200 °C, all ambipolar RGO FET remain ambipolar with increased hole and electron mobilities, while 60 % of the p-type RGO devices are transformed to ambipolar. The maximum hole and electron mobilities of the devices are 4.0 and 1.5 $cm^2 V^{-1} s^{-1}$, respectively.

Moreover, GO can be further used to construct organic FET [21]. As shown in Fig. 4.2a, the GO-based organic FET are fabricated on thermally grown 110 nm thick SiO_2 substrate using heavily doped n-type Si as the gate electrode. The GO nanosheets are then well-dispersed on SiO_2 substrate as a charge-trapping layer, which is spin-coated at 3,000 rpm for 40 s. In the following, poly(methyl methacrylate) (PMMA) is dissolved in toluene at a concentration of 10 mg/ml, spin-coated at 3,000 rpm for 40 s, and annealed on a hotplate at 120 °C for 15 min in a nitrogen-filled glove box, which is used as a charge-tunneling layer. Finally, 50 nm thick gold source and drain electrodes are deposited with a shadow mask by thermal evaporation at a pressure of $\sim 10^{-7}$ Torr. Figure 4.2b plots the transfer curves of the prepared GO-based organic FETs, which displays large gate bias dependent hysteresis with threshold voltage shifts over 20 V. After writing and erasing, the stored data are well maintained with ON/OFF ratio in the order of 10^2 for 10^4 s, as shown in Fig. 4.2c.

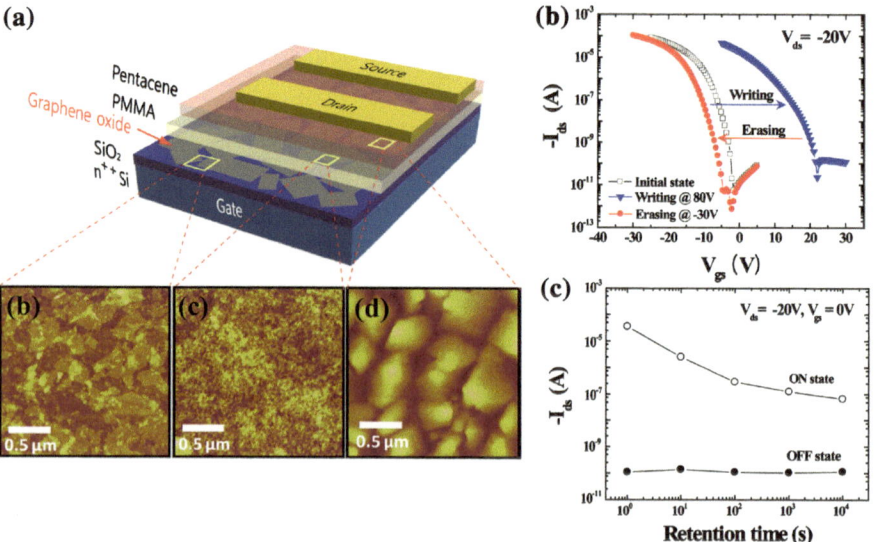

Fig. 4.2 a Device schematics of GO nanosheets based GO-based organic FET with n^{++} Si/SiO$_2$/GO nanosheets/PMMA/pentacene/Au source-drain stacks. **b** Transfer curve characteristics (I_{ds} vs. V_{gs}) of GO-based organic FET, according to the different writing ($V_{gs} = 80$ V)/erasing ($V_{gs} = -30$ V) biases at drain bias (V_{ds}) of -20 V. **c** Retention characteristics of GO-based organic FET. After writing or erasing process, drain current (I_{ds}) was read for 10^4 s at drain bias (V_{ds}) of -20 V and gate bias (V_{gs}) of 0 V condition. Reprinted with permission from Ref. [21]. Copyright (2010) AIP Publishing LLC

The adjustable thickness of GO film also enables its application in the thin film transistors. Using a solution-based method, Eda et al. [22] were able to deposit RGO thin films with monolayer to several-layer thicknesses over large substrates. The single-layer RGO films exhibit graphene-like ambipolar transistor characteristics, whereas thicker films behave as graphite-like semimetals. Then, by depositing and reducing GO on p-Si substrates with 300 nm oxide layer as the gate electrode, they fabricated a bottom-gated thin-film transistor, as shown in Fig. 4.3a. The source and drain electrodes are individually deposited with gold with channel lengths of 21 and 210 mm. The transfer characteristics as a function of temperature for the 20 and 80 ml RGO thin films are depicted in Fig. 4.3b, c, where the low-temperature measurements exhibit ambipolar characteristics, comparable to that of graphene. For the ambient characteristics, the hole and electron mobilities of the devices are ~1 and ~0.2 $cm^2 V^{-1} s^{-1}$, respectively. Afterwards, Eda and Chhowalla designed a RGO-composite thin-film transistor using RGO as the filler and polystyrene (PS) as the host [23], as shown in Fig. 4.3d. The transistor exhibits ambipolar field effect characteristics (Fig. 4.3e), suggesting transport via percolation among RGO in the insulating PS matrix. The ambipolar characteristics are comparable to those of RGO films. Hence the influence of PS in charge conduction and gate capacitance is minor. Unlike RGO films, the composite

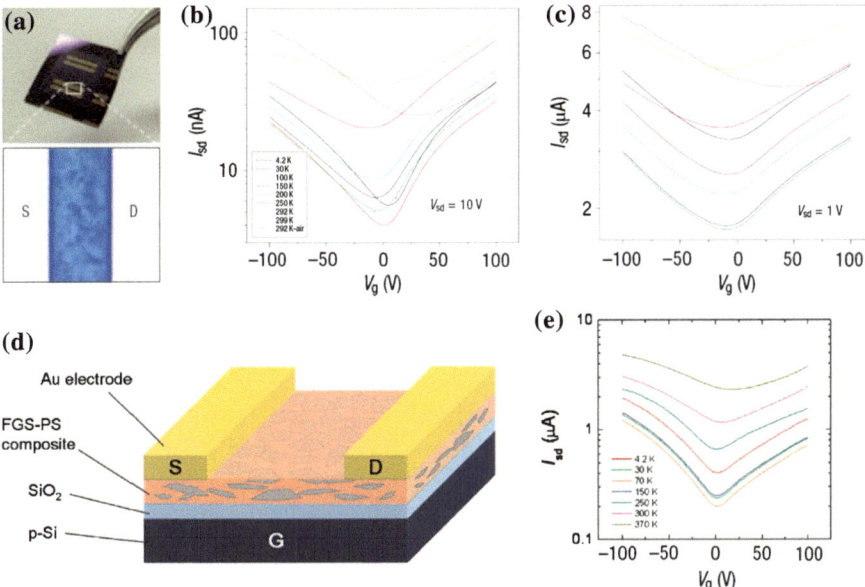

Fig. 4.3 a Optical micrograph of the GO thin-film transistor; source–drain current as a function of gate voltage for 20 ml (**b**) and 80 ml (**c**) films measured at different temperatures. Reprinted with permission from Ref. [22]. Copyright (2008) Nature Publishing Group. **d** Schematic of RGO-PS composite thin-film transistor; **e** transfer characteristics at various temperatures in log scale ($V_{sd} = 1$ V). Reprinted with permission from Ref. [23]. Copyright (2009) American Chemical Society

devices are only weakly sensitive to unintentional ambient doping, demonstrating that the majority of RGO responsible for the carrier transport are embedded within the PS. Therefore, the insulating PS matrix provides structural integrity and air stability while minimally interfering with the electrical properties of the RGO network [12].

In addition, GO-based composites are anticipated to function as efficient field emitters. By putting GO sheets on Ni nanotip arrays, high density of sharp protrusions within the sheets can be produced, which further lead to efficient and stable field emission with low turn-on fields [24]. The field enhancement occurring at protrusions in these GO sheets results in a turn-on electric field of only 5×10^5 V/m at 1 μA emission current. Moreover, the emission current density can reach 1 mA/cm^2 at 10^6 V/m, which is critical for practically viable field emission applications. On the other hand, Eda et al. [25] reported improved field enhancement of the RGO-PS composites due to atomically sharp edges. Measurement of the field-emission characteristics at different spin-coating speeds shows that the threshold field required to drive a current density of 10^{-8} A cm^{-2} is significantly lower for the 600 rpm sample (~4 V μm^{-1}) in comparison to the 2,000 rpm sample (~11 V μm^{-1}). The much lower threshold field for electron emission in the 600 rpm sample suggests significantly higher field enhancement factor.

4.1.3 Flexible Electronic Materials

Flexible electronics is an emerging and promising technology for next-generation, high-performance portable electronic devices [26]. The advance of flexible electronics demands the development of thin, lightweight, and flexible electrode materials. In 2008, Eda et al. [22] reported a solution-based method for uniform and controllable deposition of RGO thin films with thicknesses ranging from one to several layers. The RGO films are very mechanical flexible, see Fig. 4.4a. The sheet resistance value of RGO can be down to 43 kΩ/square and the hole and electron mobilities can reach ~1 and ~0.2 cm^2 V^{-1} s^{-1}, respectively. Therefore, the RGO thin films can be used for flexible electronic materials.

Lee et al. [27] were able to integrate GO platelets by self-assembling into mechanically flexible, macroporous three-dimensional (3-D) carbon films with tunable porous morphologies. Meanwhile, the electrical properties of GO can be significantly enhanced through pyrolysis or nitrogen doping process, as indicated from the I-V curves in Fig. 4.4b). During the pyrolysis of grafted polymers, the surface resistance of the macroporous RGO film reduces to ~128.2 Ω. Nitrogen doping will further decrease the sheet resistances to 13.4 Ω, corresponding to an electrical conductivity of 649 S/cm where the film thickness is 1.15 μm.

Jeong et al. [28] constructed a nonvolatile flexible memory device on the basis of GO thin film. The device has a trilayer cross-point structure composed of Al/ GO film/Al sheets deposited on flexible polyethersulfone (PES) substrate. This flexible memory device shows reliable and reproducible bipolar resistive switching

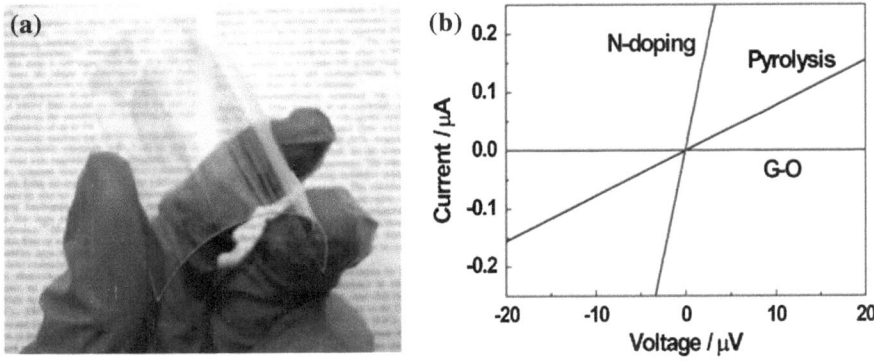

Fig. 4.4 a Photograph of transparent GO thin films on plastic substrates. Reprinted with permission from Ref. [22]. Copyright (2008) Nature Publishing Group. **b** I-V curves of GO and macroporous RGO films before and after thermal treatment. Reprinted with permission from Ref. [27]. Copyright (2010) Wiley-VCH Verlag GmbH & Co. KGaA, Weinheim

with an ON/OFF ratio of ~100 and a switching voltage of ~2.5 V. Both ON and OFF states are stable for more than 10^5 s after removal of the external voltage stimulus. The memory performance can be retained over ~100 cycles without degradation.

In addition, some all-GO-based flexible electronic devices have been fabricated. He et al. [29] fabricated the all-RGO thin film flexible transistors. The transistor consists of solution-processed RGO electrodes and a micropatterned RGO channel. Bending test shows that the all-RGO-based transistor is very flexible. Its resistance only degrades by ~1 % after the initial 200 time bending cycles and then becomes very stable without obvious resistance change even up to 5,000 bending cycles at the bend radius of 4 mm. The initial resistance drops may originate from the rearrangement of RGO sheets, which in turn enhances the intimate stacking of RGO layers. Such all-RGO-based devices exhibit good performance with ON/OFF ratio of 3.8 and Dirac point of +0.4 V in the electrolyte-gating. Moreover, the all-RGO transistor shows good sensitivity and selectivity, which can be used to detect protein (fibronectin) as low as 0.5 nM.

Liu et al. [30] also reported a nonvolatile memory device completely composed of all-RGO films. In the device, both the top and bottom electrodes are made of highly reduced GO films obtained by the high-temperature annealing of GO, whereas the active material is made of lightly reduced GO obtained by low-temperature annealing and then light irradiation of GO, as shown in Fig. 4.5a, b, respectively. The fabricated diode shows the electrical bistability and nonvolatile memory effect with a current ON/OFF ratio of 10^2. The memory device exhibits strong retention ability with a retention time as long as 10^3 s under ambient conditions and its ON/OFF ratio shows no significant variation. In particular, this all-RGO-based memory exhibits excellent flexibly with desirable long life time under retention and bending tests, as shown in

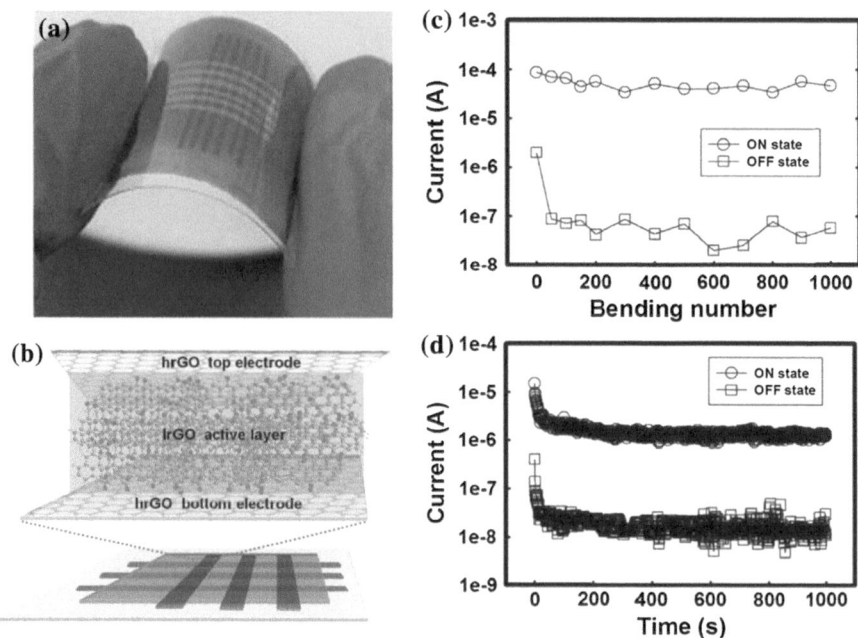

Fig. 4.5 **a** Photograph of the flexible memory device with configuration of hrGO/lrGO/hrGO and its schematic illustration (**b**). **c** Bending experiment of the all RGO-based memory device at a voltage of 2 V. **d** Retention test of the memory device at a reading voltage of 1 V in ambient conditions. Reprinted with permission from Ref. [30]. Copyright (2013) Wiley-VCH Verlag GmbH & Co. KGaA, Weinheim

Fig. 4.5c. After repetitive bending and relaxing of this device at a voltage of 2 V with a tensile strain of about 2.9 %, there is no noticeable electrical degradation in both the ON and OFF states up to 1,000 iterations (see Fig. 4.5d), which indicates a high mechanical endurance of the fabricated device. The high flexibility of the device has been mainly attributed to three factors: (1) the thermal insensitivity of lightly reduced GO at ambient conditions, (2) the electrical stability of flexible highly reduced GO electrode, (3) the compact contact between highly reduced GO (hrGO) and lightly reduced GO (lrGO). The thermal insensitivity of lightly reduced GO is originated from the low-temperature annealing, which can partially remove the oxygen-containing groups of GO and make the obtained GO less hydrophilic, resulting in enhanced film strength and better contact with the top and bottom electrodes. The electrical stability of RGO film is due to its graphene-like properties since graphene film can keep its resistance stable in both the longitude and transverse direction under stretching of up to 11 % [31]. Furthermore, the strong π–π interaction between highly reduced GO and lightly reduced GO films enables their compact contact, leading to a stable electrical contact in the hrGO/lrGO/hrGO layers.

4.1.4 Electrical Sensors

By changing the oxidation degree or adsorbing molecules, the conductance of GO may vary, which enables GO promising electrical sensors [12, 32]. Robinson et al. [33] demonstrated that the RGO networks can be used as active materials for high-performance molecular sensors owing to the change of electric conductance after molecular adsorption. Depending on the type and density of available sites, molecules may bind to RGO at different sites with different binding energies. Generally, there are two kinds of binding sites in the RGO. The first kind of site locates in graphitic sp^2 domains with weak interaction. Molecules weakly adsorbed onto RGO via dispersive forces can desorb with thermal energy at room temperature. Therefore, conductance changes associated with weakly bound molecules are recoverable upon evacuation. Another kind of binding sites are vacancies and oxygen functional groups, which interact stronger with molecules with binding energies of at least several hundred meV. In this case, conductance changes associated with molecular adsorption are non-recoverable. The GO-based sensors have low frequency noise that is orders of magnitude lower than that of SWNT-based sensors [34]. Moreover, the frequency noise of GO-based sensors is sensitive to both hydrazine hydrate exposure time and film thickness [33]. The suppressed frequency noise of RGO films can be attributed to charged impurity screening, as proposed by Lin and Avouris [35].

Similarly, He et al. [36] found that the conductance of RGO-based device changes with the concentration of dopamine molecule. As shown in Fig. 4.6, at the gate voltage (V_g) of –0.6 V, the conductance of RGO-PET device (RGO films on flexible polyethylene terephthalate (PET) substrate) increases with increasing

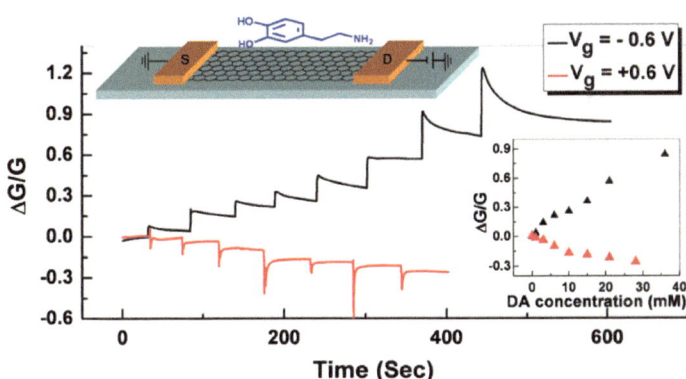

Fig. 4.6 Detection of dopamine on RGO-PET device at $V_g = \pm0.6$ V. Each step represents the gradual addition of dopamine with concentration increasing from 1 to 8 mM. *Inset* change of conductance versus dopamine concentration. The distance between the drain and source electrodes in the RGO-PET device is fixed at 1 cm. Reprinted with permission from Ref. [36]. Copyright (2010) American Chemical Society

concentration of dopamine. However, at the V_g of +0.6 V, the conductance of RGO-PET device will decrease with increasing concentration of dopamine. This is because the RGO device has a Dirac voltage of +0.2 V. Above it, the device operates at n-type region; below it, the device operates at p-type region. After removing dopamine by thorough perfusion, the current returns to its initial level and the device can be reliably used for many times. In addition, GO-based devices have been employed to detect other molecules such as H_2O, NO_2 and NH_3 by the change of electrical conductance, as reviewed in Ref. [37].

4.2 GO for Optics

4.2.1 Fluorescence Quenching

Due to the heterogeneous chemical, atomic and electronic structures, GO can quench fluorescence despite itself being fluorescent. As reported previously, graphitic carbon can quench fluorescence from dye molecules adsorbed on the surfaces evidenced by the Raman spectroscopy [38]. Similarly, GO with sp^2 carbon network domains also allows quenching of nearby fluorescent species such as dyes, conjugated polymers and quantum dots [3]. This fluorescence quenching effect can be attributed to fluorescence (or Förster) resonance energy transfer, or non-radiative dipole–dipole coupling, between the fluorescent species and GO [39–41]. On the other hand, the quenching efficiency of GO can be also enhanced by several methods. For instance, reduction of GO will significantly improve its fluorescence quenching efficiency [39]. Another way is to enhance the π–π stacking between the fluorescent species and GO [42].

Based on the quenching effect of GO, a so-called fluorescence quenching microscopy (FQM) technique has been developed [39, 43], allowing the visualization of morphological characteristics of individual GO and RGO sheets on arbitrary substrates or inside liquids. Under FQM characterization, GO or RGO provides dark contrast with the surrounding fluorescent medium, which significantly enhances the contrast of GO and RGO compared with those with the conventional optical imaging techniques [3]. The enhanced contrast can be ascribed to a removal of the signal from the excitation source via a low-pass optical filter. Using FQM, Kim et al. [39] was able to easily visualize the morphologies of GO or RGO layers coated with dye molecules, as illustrated in Fig. 4.7. Due to different fluorescence quenching efficiency, it is also easy to distinguish GO and RGO layers. In addition, the large effective remote quenching distance of GO is about 20 nm, close to the theoretically predicted value (~30 nm) for the pristine graphene [44, 45].

Another application of GO based on the fluorescence quenching effect is to serve as fluorescence sensors, which is very useful in biosensing. Since GO contains different kinds of functional groups, ionic groups, such as O^- and COO^-, allow electrostatic interactions between the charged biomolecules and GO.

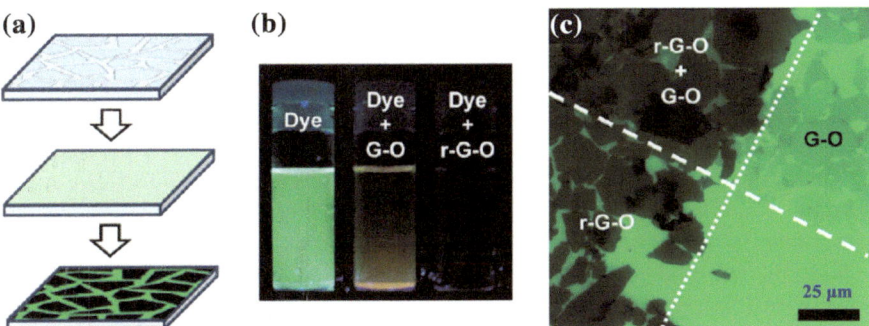

Fig. 4.7 **a** A dye is coated to visualize GO and RGO by FQM. **b** A camera image showing fluorescence quenching effect of GO and RGO in a fluorescein solution. **c** A glass coverslip deposited with GO and RGO samples showing different quenching efficiency. Reprinted with permission from Ref. [39]. Copyright (2010) American Chemical Society

For instance, GO can bind dye-labeled single-stranded DNA and completely quench the fluorescence of the dye through a strong noncovalent π–π stacking interactions [46]. However, the presence of a target will lead to conformation change of the dye-labeled DNA, which further disturbs the interaction between the dye-labeled DNA and the GO. Interruption of the interaction will release the dye-labeled DNA from the GO, resulting in restoration of dye fluorescence. As a consequence, this strategy results in a fluorescence-enhanced detection that is sensitive and selective to the target molecule. Due to the high planar surface and universal quenching capabilities of GO, this mechanism can be extended to a multiplexed (multicolor) DNA detection [47] or a multiplexed detection of different targets (DNA, proteins, metal ions, etc.) [48]. The other type of GO-based biosensors will be discussed in Chap. 7.

4.2.2 SERS

SERS is a surface-sensitive technique that enhances Raman scattering by molecules adsorbed on rough metal surfaces or by nanostructures, which can be used to detect single molecules due to its high sensitivity. Huang et al. [49] demonstrated that the Au nanoparticles (AuNPs) dispersed on the GO and RGO supports exhibit better SERS performance compared with the metal nanoparticles alone. The SERS intensity of the p-aminothiophenol on the Au-GO composites is much stronger than on the AuNPs with the same concentration of p-aminothiophenol since the AuNPs on the Si substrate are isolated with few SERS "hot spots", while the AuNPs on the GO nanoplatelets are aggregated, leading to a coupled electromagnetic effect and further resulting in a significant enhanced Raman scattering of the p-aminothiophenol molecules.

Similarly, the GO/Ag nanoparticle hybrids also show better SERS performance than the Ag nanoparticles (AgNPs) alone [50]. The composite of GO, poly(diallyldimethyl ammonium chloride) (PDDA) and AgNPs, which is shown in Fig. 4.8a, exhibit strong SERS activity to folic acid due to the electrostatic interaction between the folic acid molecules and the GO. Moreover, the detection sensitivity depends on the concentration of graphitic carbon in GO/PDDA/AgNP solution. The SERS spectrum of folic acid ($1,595$ cm^{-1}) is clear enough and background of GO (D band at $1,355$ cm^{-1}) is ignorable, as shown in Fig. 4.8b. Besides, the minimum detected concentration of the folic acid in water is as low as 9 nM (Fig. 4.8c), and the calibration curve shows a linear relation within a linear response range from 9 to 180 nM (Fig. 4.8d).

On the other hand, chemical enhancement of SERS can be tailored by controlling the chemical reduction of GO nanosheets. Yu et al. [51] systematically examined the SERS of Rhodamine B (RhB) molecules on CRGO as a function

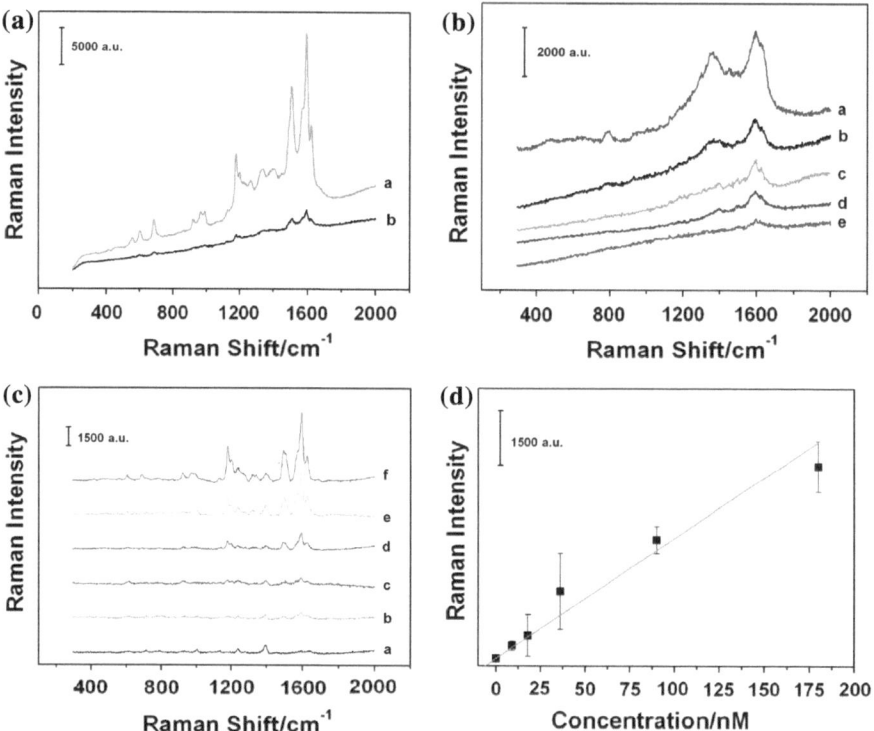

Fig. 4.8 **a** SERS spectra of 10^{-4} M folic acid on GO/PDDA/AgNPs (*a*) and AgNP colloid (*b*); **b** SERS spectra of 9 nM folic acid obtained in the GO/PDDA/AgNP solutions with 0.02 mg/mL (*a*), 0.01 mg/mL (*b*), 0.004 mg/mL (*c*), 0.0025 mg/mL (*d*), and 0.002 mg/mL (*e*) graphitic carbon; **c** SERS spectra of different concentrations of folic acid in water: blank (*a*), 9 nM (*b*), 18 nM (*c*), 36 nM (*d*), 90 nM (*e*), and 180 nM (*f*); **d** SERS dilution series of folic acid in water based on the peak located at $1,595$ cm^{-1}. Reprinted with permission from Ref. [50]. Copyright (2011) American Chemical Society

of reduction time. Compared with the mechanically exfoliated graphene, mildly reduced GO nanosheets can significantly increase the chemical enhancement of the main peaks by up to one order of magnitude for the adsorbed RhB molecules. The observed enhancement factors can be as large as ~10^3 and show distinct dependence on the reduction time of GO, indicating that the chemical enhancement can be steadily controlled by specific chemical groups. The strong enhancement of SERS might be associated with aromatic C–C bonds, suggesting that the aromatic rings in RhB are closely interacting with the substrate. The π-π stacking and the lone pair electrons in the oxygen-containing groups between the RhB molecule and the mildly reduced GO are the major driving force for the large Raman enhancement.

4.2.3 Nonlinear Optical Materials

Nonlinear optical materials, including optical-limiting materials and saturable absorbers, are of great importance to modulate the laser intensity. Optical-limiting materials display a decreased transmittance at high input laser intensity and can be used to protect eyes and sensitive instruments from laser-induced damage. On the contrary, saturable absorbers present an increased transmittance at high input laser intensity, which are useful for pulse compression, mode locking, and Q-switching [52]. Previously, it was reported that graphene shows good nonlinear optical properties and can serve in ultrafast lasers as saturable absorbers [53, 54]. Since GO has a tunable energy gap under different degrees of oxidation, it should even surpass the performance of graphene for such applications in principle.

Generally, GO shows a better optical limiting response than that of the benchmark material, i.e. C_{60}. Jiang et al. [52] investigated the nonlinear optical properties of GO thin films on glass and plastic substrates. The as-prepared GO films exhibit excellent broadband optical limiting behaviors (see Fig. 4.9a). Particularly, their optical limiting activity could be significantly enhanced upon partial reduction by using laser irradiation or chemical reduction methods, as shown in Fig. 4.9a–c. For femtosecond laser pulses at 400 nm, the laser-induced reduction of GO results in enhancement of effective two-photon absorption coefficient by up to ~19 times; for femtosecond laser pulses at 800 nm, enhancement of effective two- and three-photon absorption coefficients by ~12 and ~14.5 times are achieved, respectively. Besides, the optical limiting thresholds of partially reduced GO films are much lower than those of previously reported materials, indicating excellent applications of the highly reduced GO films as saturable absorbers. In addition to reduction, the optical limiting properties of GO can be also improved by linking it with nonlinear optical molecules, such as porphyrin [55], oligothiophene [56] and C_{60} fullerene [57]. Compared with benchmark fullerene materials, hybrid GO–dye materials present superior optical limiting effects. This can be ascribed to a combination of optical limiting mechanisms, including nonlinear

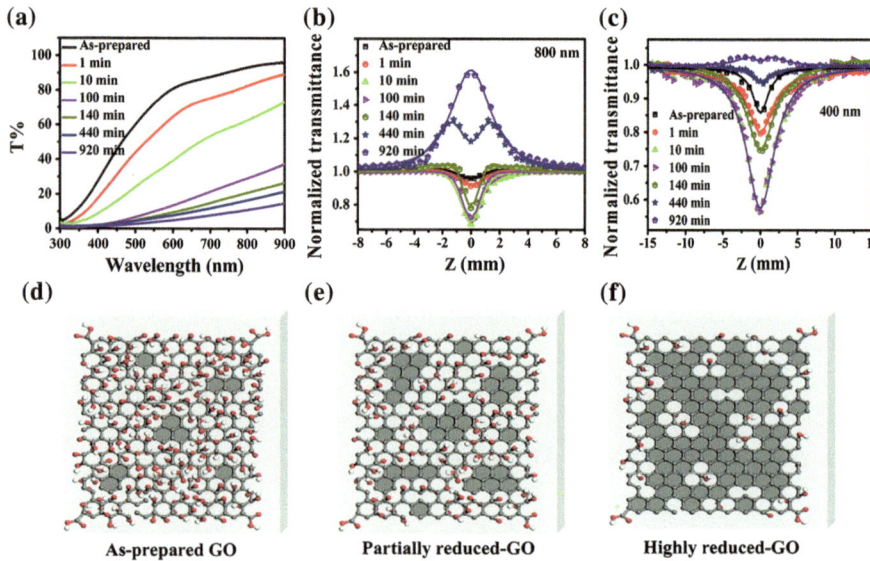

Fig. 4.9 Linear transmittance (**a**) and Z-scan measurement results at 800 nm (fluence of 25 mJ cm^{-2} at the focal point) (**b**) and at 400 nm (fluence of 0.6 mJ cm^{-2} at the focal point) (**c**) of a GO film upon exposure to hydrazine vapor for different periods of time. Schematic structures of as-prepared (**d**), partially reduced (**e**), and highly reduced (**f**) GO at different stages of reduction. Reprinted with permission from Ref. [52]. Copyright (2012) American Chemical Society

optical absorption and scattering, as well as photo-induced electron or energy transfer in GO–organic hybrids [3].

Usually, GO is not suitable for broadband saturable absorber in laser cavities for ultrafast pulse generation since the carbon atoms bonded with oxygen groups are sp^3 hybridized and disrupt the sp^2 conjugation of the hexagonal graphene lattice, which further destroy the linear dispersion of the Dirac electrons and influence the unique optical properties of graphene [58]. However, employing femtosecond pump-probe and Z-scan techniques, Zhao et al. [59] demonstrated that few-layered GO films show a fast energy relaxation of hot carriers and strong saturable absorption. The fast carrier relaxation and large saturable absorption of FGO in N,N-dimethylmethanamide (DMF) solution indicate that oxidation mainly occurs at the edge areas and has ignorable effect on ultrafast dynamics and optical nonlinearities. Large fraction of sp^2 carbon atoms exists inside the few-layered GO sheet. Later, Sobon et al. [60] comprehensive studied GO and RGO based saturable absorbers for mode-locking of Er-doped fiber lasers, as shown in Fig. 4.10a. No significant difference in the laser performance between the GO and RGO based saturable absorbers is found. Both saturable absorbers provide stable and mode-locked operation with 390 fs soliton pulses and more than 9 nm optical bandwidth at 1,560 nm center wavelength. The RGO-based absorber has a relatively high modulation depth of 21 % than that of the GO-based absorber (18 %).

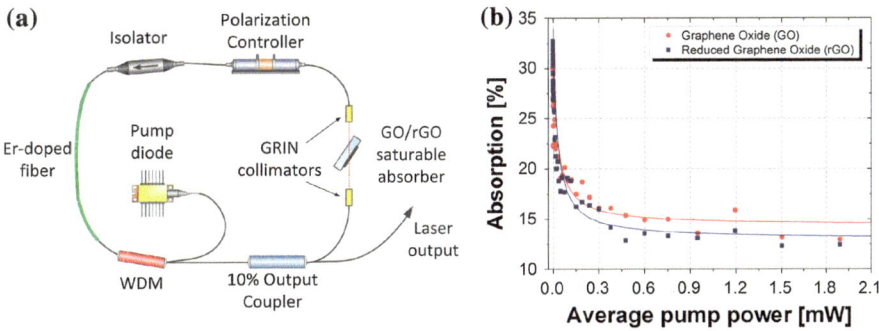

Fig. 4.10 **a** Experimental setup of the mode-locked laser; and **b** power-dependent transmission of the GO and RGO. Reprinted with permission from Ref. [60]. Copyright (2012) OSA

Table 4.1 Summary of the laser parameters with GO and RGO based absorber

Parameter	Value	
	GO	RGO
FWHM bandwidth (nm)	9.3	9.2
Pulse duration (fs)	390	390
TBP	0.448	0.442
Pulse energy (pJ)	33.7	29.8
Peak power (W)	86.4	76.4
Soliton order N	0.79	0.75
RF SNR (dB)	60	60
Pump power (mW)	92	82
Output power (mW)	1.96	1.68
Center wavelength (nm)	1,558	1,559

FWHM full width at half maximum, *TBP* time-bandwidth products; *RF* is radio frequency, and *SNR* signal to noise ratio. Reprinted with permission from Ref. [60]. Copyright (2012) OSA

However, both of them show low non-saturable loss around 15 % (see Fig. 4.10b). Considering of the cost of production, GO seems to be a better saturable absorber than RGO. Table 4.1 summarizes the laser parameters with GO and RGO based absorber.

4.3   GO for Optoelectronics

4.3.1   Photovoltaic Devices

Based on the transparent and conducting properties, GO films can be used in opto-electronic devices, such as photovoltaic [4, 11, 61] and light-emitting [6] devices. Generally, for these optoelectronic devices, the appropriate band alignment of each component and the electrical resistance of carrier transport layers are of key importance to determine the efficiency of carrier collection and carrier injection. Previously, Shin et al. [62] showed that the work function of RGO varies between 4.2 and 4.4 eV (depending on the C:O ratio), which almost coincides with that of ITO (4.4–4.5 eV) [63]. Thus, RGO should be a potential replacement of ITO for photovoltaic or LED devices. In 2008, Wang et al. [11] investigated a dye-sensitized solar cell (DSSC) with RGO as the transparent electrode and found that as the dye absorbs sun light and becomes excited, the electrons are injected into the conduction band of TiO$_2$ and transported to the RGO electrode. Similarly, holes are transported through the hole-transport layer and collected by the Au cathode. For comparison, a cell with a fluorine tin oxide (FTO) electrode was also fabricated and evaluated with the same procedure by Wang et al. [11]. Compared with the FTO-based cell, RGO-based device has lower short-circuit current, which can be attributed to the higher sheet resistance and lower transmittance of the RGO.

Similar results have also been reported for organic photovoltaic (OPV) devices with RGO electrodes [17, 64]. Owing to the higher sheet resistance, the RGO-based OPV solar cells show lower short-circuit current and fill factor than those of control device on ITO. By transferring CRGO onto PET substrates to act as transparent and conductive electrodes, Yin et al. [65] fabricated flexible GO-based OPV devices, as displayed in Fig. 4.11. When the optical transmittance of RGO is larger than 65 %, the device performance mainly relies on the charge transport efficiency through RGO electrodes, but not sensitive to the transmittance. Therefore, the current density of devices can be enhanced by increasing the RGO thickness to lower the sheet resistance, which further improves the overall power conversion efficiency (η), even if the transmittance of RGO film decreases.

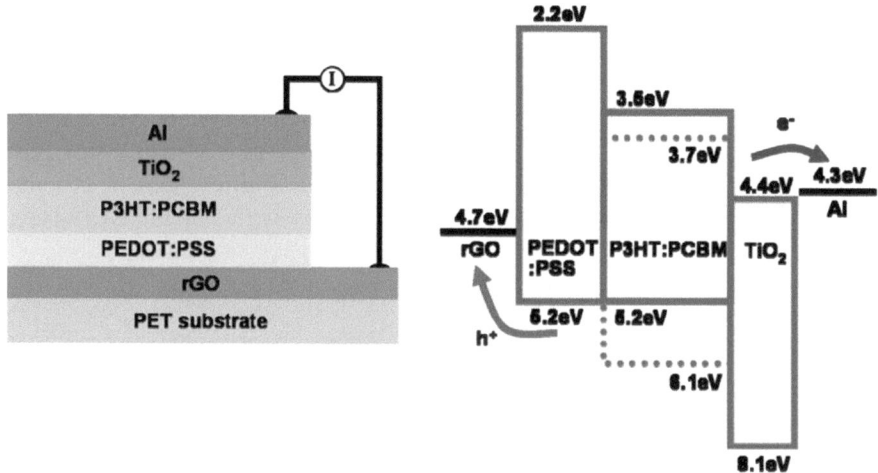

Fig. 4.11 Schematic representation of the layer structure (*left*) and energy level (*right*) for the OPV device with RGO as the transparent electrode. Reprinted with permission from Ref. [65]. Copyright (2010) American Chemical Society

Table 4.2 Photovoltaic performance parameters for OPV devices fabricated with either PEDOT:PSS or GO interfacial layer

IFL	J_{sc} (mA/cm^2)	σ_{Jsc}	V_{oc} (V)	σ_{Voc}	FF (%)	σ_{FF}	η (%)	σ_{η}
PEDOT:PSS	−14.55	0.0683	0.741	0.0019	68.10	0.6179	7.46	0.0857
GO	−15.21	0.0239	0.716	0.0021	67.70	0.3627	7.39	0.0497

Here, J_{sc} average short circuit current, V_{oc} open circuit voltage, FF fill factor, σ are the derived standard deviations. Reprinted with permission from Ref. [66]. Copyright (2011) American Chemical Society

It was found that the highest η obtained in the flexible RGO/PET-based OPV device is 0.78 %.

Murray et al. [66] also demonstrated that GO could be good OPV material, which can be used to replace the standard electron-blocking layer, poly-3,4-ethylenedioxy-thiophene:poly(styrene sulfonate) (PEDOT:PSS) and significantly enhances device durability, as presented in Table 4.2. Such GO-based OPV devices exhibit η as high as ~7.5 % while providing a five times enhancement in thermal aging lifetime and a twenty times enhancement in humid ambient lifetime compared with the analogous PEDOT:PSS-based devices.

In addition, Stratakis et al. [67] reported a photochemical method to improve the efficiency of GO-based OPV by tuning the work function of GO hole transporting layers. The GO films are doped with Cl via ultraviolet laser irradiation in the presence of a Cl_2 precursor gas. By changing the laser exposure time, the work function of the GO–Cl layers can be tailored from 4.9 eV to a maximum value of 5.23 eV by controlling the doping and reduction levels. As a result, high efficiency poly(2,7-carbazole) derivative (PCDTBT):fullerene derivative ($PC_{71}BM$) based OPVs with GO–Cl as the hole transporting layer achieve a power conversion efficiency η of 6.56 %, which is 17.35 and 19.48 % higher than that of the pristine GO and PEDOT:PSS based OPV devices, respectively. Such performance enhancement is attributed to more efficient hole transportation due to the energy level matching between the GO–Cl and the polymer donor.

4.3.2 Light-Emitting Devices

Another important optoelectronic application of GO is to serve as light-emitting materials. By surface functionalization of GO with anthryl moieties using 2-aminoanthracene diazonium salts as grafting agents, Lu et al. [68] achieved the blue-emitting GO nanosheets. Different from the fluorescence quenching of GO to quantum dots in their hybrid materials, the anthryl moieties functionalized graphene oxide (GO-A) exhibits strong photoluminescence. The as-synthesized GO hybrid composites show strong blue photoluminescence centered at ~400 nm, which is distinctly different from the cyan emission of monomeric 2-aminoanthracene centered at ~491 nm. This large blue shift of the luminescence about ~91 nm

Table 4.3 Device performance of polymer LEDs with different hole transport layers

Device configuration	Maximum luminance (cd/m^2)	Maximum luminous efficiency (cd/A)	Maximum power efficiency (lm/W)	Maximum EQE (%)	Turn-on voltage (V)
ITO/SY/LiF/Al	700 (at 16.0 V)	1.4 (at 8.4 V)	0.6 (at 6.6 V)	0.6 (at 8.4 V)	2.8
ITO/PEDOT:PSS/SY/LiF/Al	33,800 (at 12.6 V)	8.7 (at 9.6 V)	3.9 (at 5.2 V)	3.5 (at 9.2 V)	1.8
ITO/GO[2.0 nm]/SY/LiF/Al	31,400 (at 12.4 V)	8.8 (at 9.4 V)	4.2 (at 5.0 V)	3.3 (at 8.2 V)	1.8
ITO/GO[2.6 nm]/SY/LiF/Al	35,100 (at 12.0 V)	14.3 (at 8.6 V)	6.6 (at 5.4 V)	5.0 (at 8.4 V)	1.8
ITO/GO[4.3 nm]/SY/LiF/Al	39,000 (at 10.8 V)	19.1 (at 6.8 V)	11.0 (at 4.4 V)	6.7 (at 6.8 V)	1.8
ITO/GO[5.2 nm]/SY/LiF/Al	28,500 (at 11.2 V)	13.9 (at 7.4 V)	8.6 (at 4.0 V)	5.0 (at 7.4 V)	1.8
ITO/rGO[4.3 nm]/SY/LiF/Al	8,300 (at 13.0 V)	5.0 (at 8.6 V)	2.0 (at 6.2 V)	1.8 (at 8.6 V)	1.8

for the functionalized GO hybrids can be partly ascribed to the rigid chemical environment with anthryl moieties chemically bonded onto GO surface. Therefore, GO can be made luminescent by chemical bonding fluorescence molecules on the stable GO frameworks, which is useful in the fabrications of optoelectronic devices and fluorescent sensors.

Using the GO as hole transport layer between the ITO and the active polymer instead of PEDOT:PSS, Lee et al. [69] fabricated polymer light-emitting diodes (LEDs) with enhanced device efficiency. Compared with the devices with PEDOT:PSS and GO alone, the polymer LEDs with a GO interlayer shows better performance with the maximum luminance of 39,000 Cd/m^2, the maximum efficiency level of 19.1 Cd/A (at 6.8 V), and the maximum power efficiency of 11.0 lm/W (at 4.4 V). The luminous efficiencies with GO are approximately 2.2-times higher than that with PEDOT:PSS and 4-times higher than that with GO. The improved efficiency of polymer LEDs is ascribed to the electron-blocking behaviors of the GO layer since GO layer has a wide band gap and can block electron transport from an emissive polymer to an ITO anode while reducing the exciton quenching between the GO and the active layer. Furthermore, thickness of the GO layer is an important factor to determine the device performance, as summarized in Table 4.3. The optimal GO layer thickness is 4.3 nm, which gives the maximum luminance, luminous efficiencies, and power efficiency. In addition to serving as interfacial layer [69–71], GO can be also used as the electrode material after reduction. The patterned RGO films can exhibit good conductivity and serve as electrodes in organic light-emitting devices (OLED) [72].

References

1. Chen, J.H., Cullen, W.G., Jang, C., Fuhrer, M.S., Williams, E.D.: Phys. Rev. Lett. **102**, 236805 (2009)
2. Kim, K., Park, H.J., Woo, B.C., Kim, K.J., Kim, G.T., Yun, W.S.: Nano Lett. **8**, 3092–3096 (2008)
3. Loh, K.P., Bao, Q.L., Eda, G., Chhowalla, M.: Nat. Chem. **2**, 1015–1024 (2010)
4. Eda, G., Lin, Y.Y., Miller, S., Chen, C.W., Su, W.F., Chhowalla, M.: Appl. Phys. Lett. **92**, 233305 (2008)
5. Wu, J., Becerril, H.A., Bao, Z., Liu, Z., Chen, Y., Peumans, P.: Appl. Phys. Lett. **92**, 263302–263303 (2008)
6. Wu, J., Agrawal, M., Becerril, H.A., Bao, Z., Liu, Z., Chen, Y., Peumans, P.: ACS Nano **4**, 43–48 (2009)
7. Wassei, J.K., Kaner, R.B.: Mater. Today **13**, 52–59 (2010)
8. Becerril, H.A., Mao, J., Liu, Z., Stoltenberg, R.M., Bao, Z., Chen, Y.: ACS Nano **2**, 463–470 (2008)
9. Ning, J., Wang, J., Li, X., Qiu, T., Luo, B., Hao, L., Liang, M., Wang, B., Zhi, L.: J. Mater. Chem. A **2**, 10969–10973 (2014)
10. Meenakshi, P., Karthick, R., Selvaraj, M., Ramu, S.: Sol. Energy Mater. Sol. Cells **128**, 264–269 (2014)
11. Wang, X., Zhi, L., Müllen, K.: Nano Lett. **8**, 323–327 (2008)
12. Eda, G., Chhowalla, M.: Adv. Mater. **22**, 2392–2415 (2010)
13. Li, J., Hu, L., Wang, L., Zhou, Y., Grüner, G., Marks, T.J.: Nano Lett. **6**, 2472–2477 (2006)
14. Zhu, Y., Murali, S., Cai, W., Li, X., Suk, J.W., Potts, J.R., Ruoff, R.S.: Adv. Mater. **22**, 3906–3924 (2010)
15. Novoselov, K.S., Geim, A.K., Morozov, S.V., Jiang, D., Zhang, Y., Dubonos, S.V., Grigorieva, I.V., Firsov, A.A.: Science **306**, 666–669 (2004)
16. Han, M.Y., Özyilmaz, B., Zhang, Y., Kim, P.: Phys. Rev. Lett. **98**, 206805 (2007)
17. Chen, Z., Lin, Y.M., Rooks, M.J., Avouris, P.: Physica E **40**, 228–232 (2007)
18. Li, X., Wang, X., Zhang, L., Lee, S., Dai, H.: Science **319**, 1229–1232 (2008)
19. Yamaguchi, H., Murakami, K., Eda, G., Fujita, T., Guan, P., Wang, W., Gong, C., Boisse, J., Miller, S., Acik, M.: ACS Nano **5**, 4945–4952 (2011)
20. Joung, D., Chunder, A., Zhai, L., Khondaker, S.I.: Nanotechnology **21**, 165202 (2010)
21. Kim, T.W., Gao, Y., Acton, O., Yip, H.L., Ma, H., Chen, H., Jen, A.K.Y.: Appl. Phys. Lett. **97**, 023310 (2010)
22. Eda, G., Fanchini, G., Chhowalla, M.: Nat. Nanotechnol. **3**, 270–274 (2008)
23. Eda, G., Chhowalla, M.: Nano Lett. **9**, 814–818 (2009)
24. Ye, D., Moussa, S., Ferguson, J.D., Baski, A.A., El-Shall, M.S.: Nano Lett. **12**, 1265–1268 (2012)
25. Eda, G., Unalan, H.E., Rupesinghe, N., Amaratunga, G.A.J., Chhowalla, M.: Appl. Phys. Lett. **93**, 233502 (2008)
26. Zhou, G., Li, F., Cheng, H.-M.: Energy Environ. Sci. **7**, 1307–1338 (2014)
27. Lee, S.H., Kim, H.W., Hwang, J.O., Lee, W.J., Kwon, J., Bielawski, C.W., Ruoff, R.S., Kim, S.O.: Angew. Chem. **122**, 10282–10286 (2010)
28. Jeong, H.Y., Kim, J.Y., Kim, J.W., Hwang, J.O., Kim, J.E., Lee, J.Y., Yoon, T.H., Cho, B.J., Kim, S.O., Ruoff, R.S.: Nano Lett. **10**, 4381–4386 (2010)
29. He, Q., Wu, S., Gao, S., Cao, X., Yin, Z., Li, H., Chen, P., Zhang, H.: ACS Nano **5**, 5038–5044 (2011)
30. Liu, J., Yin, Z., Cao, X., Zhao, F., Wang, L., Huang, W., Zhang, H.: Adv. Mater. **25**, 233–238 (2013)
31. Reina, A., Jia, X., Ho, J., Nezich, D., Son, H., Bulovic, V., Dresselhaus, M.S., Kong, J.: Nano Lett. **9**, 30–35 (2009)
32. Jung, I., Dikin, D.A., Piner, R.D., Ruoff, R.S.: Nano Lett. **8**, 4283–4287 (2008)

33. Robinson, J.T., Perkins, F.K., Snow, E.S., Wei, Z., Sheehan, P.E.: Nano Lett. **8**, 3137–3140 (2008)
34. Robinson, J.A., Snow, E.S., Badescu, S.C., Reinecke, T.L., Perkins, F.K.: Nano Lett. **6**, 1747–1751 (2006)
35. Lin, Y.M., Avouris, P.: Nano Lett. **8**, 2119–2125 (2008)
36. He, Q., Sudibya, H.G., Yin, Z., Wu, S., Li, H., Boey, F., Huang, W., Chen, P., Zhang, H.: ACS Nano **4**, 3201–3208 (2010)
37. Basu, S., Bhattacharyya, P.: Sens. Actuators, B **173**, 1–21 (2012)
38. Kagan, M.R., McCreery, R.L.: Anal. Chem. **66**, 4159–4165 (1994)
39. Kim, J., Cote, L.J., Kim, F., Huang, J.: J. Am. Chem. Soc. **132**, 260–267 (2010)
40. Wang, Y., Kurunthu, D., Scott, G.W., Bardeen, C.J.: J. Phys. Chem. C **114**, 4153–4159 (2010)
41. Dong, H., Gao, W., Yan, F., Ji, H., Ju, H.: Anal. Chem. **82**, 5511–5517 (2010)
42. Liu, C., Wang, Z., Jia, H., Li, Z.: Chem. Commun. **47**, 4661–4663 (2011)
43. Kim, J., Kim, F., Huang, J.: Mater. Today **13**, 28–38 (2010)
44. Swathi, R.S., Sebastian, K.L.: J. Chem. Phys. **129**, 054703 (2008)
45. Swathi, R.S., Sebastian, K.L.: J. Chem. Phys. **130**, 086101 (2009)
46. Lu, C.H., Yang, H.H., Zhu, C.L., Chen, X., Chen, G.N.: Angew. Chem. **121**, 4879–4881 (2009)
47. He, S., Song, B., Li, D., Zhu, C., Qi, W., Wen, Y., Wang, L., Song, S., Fang, H., Fan, C.: Adv. Funct. Mater. **20**, 453–459 (2010)
48. Zhang, M., Yin, B.C., Tan, W., Ye, B.C.: Biosens. Bioelectron. **26**, 3260–3265 (2011)
49. Huang, J., Zhang, L., Chen, B., Ji, N., Chen, F., Zhang, Y., Zhang, Z.: Nanoscale **2**, 2733–2738 (2010)
50. Ren, W., Fang, Y., Wang, E.: ACS Nano **5**, 6425–6433 (2011)
51. Yu, X., Cai, H., Zhang, W., Li, X., Pan, N., Luo, Y., Wang, X., Hou, J.G.: ACS Nano **5**, 952–958 (2011)
52. Jiang, X.F., Polavarapu, L., Neo, S.T., Venkatesan, T., Xu, Q.H.: J. Phys. Chem. Lett. **3**, 785–790 (2012)
53. Bao, Q., Zhang, H., Wang, Y., Ni, Z., Yan, Y., Shen, Z.X., Loh, K.P., Tang, D.Y.: Adv. Funct. Mater. **19**, 3077–3083 (2009)
54. Bao, Q., Zhang, H., Yang, J.X., Wang, S., Tang, D.Y., Jose, R., Ramakrishna, S., Lim, C.T., Loh, K.P.: Adv. Funct. Mater. **20**, 782–791 (2010)
55. Xu, Y., Liu, Z., Zhang, X., Wang, Y., Tian, J., Huang, Y., Ma, Y., Zhang, X., Chen, Y.: Adv. Mater. **21**, 1275–1279 (2009)
56. Liu, Y., Zhou, J., Zhang, X., Liu, Z., Wan, X., Tian, J., Wang, T., Chen, Y.: Carbon **47**, 3113–3121 (2009)
57. Liu, Z.B., Xu, Y.F., Zhang, X.Y., Zhang, X.L., Chen, Y.S., Tian, J.G.: J. Phys. Chem. B **113**, 9681–9686 (2009)
58. Bao, Q., Zhang, H., Yang, J.X., Wang, S., Tang, D.Y., Jose, R., Ramakrishna, S., Lim, C.T., Loh, K.P.: Adv. Funct. Mater. **20**, 782–791 (2010)
59. Zhao, X., Liu, Z.B., Yan, W.B., Wu, Y., Zhang, X.L., Chen, Y., Tian, J.G.: Appl. Phys. Lett. **98**, 121905 (2011)
60. Sobon, G., Sotor, J., Jagiello, J., Kozinski, R., Zdrojek, M., Holdynski, M., Paletko, P., Boguslawski, J., Lipinska, L., Abramski, K.M.: Opt. Express **20**, 19463–19473 (2012)
61. Su, Q., Pang, S., Alijani, V., Li, C., Feng, X., Müllen, K.: Adv. Mater. **21**, 3191–3195 (2009)
62. Shin, H.J., Kim, K.K., Benayad, A., Yoon, S.M., Park, H.K., Jung, I.S., Jin, M.H., Jeong, H.K., Kim, J.M., Choi, J.Y.: Adv. Funct. Mater. **19**, 1987–1992 (2009)
63. Park, Y., Choong, V., Gao, Y., Hsieh, B.R., Tang, C.W.: Appl. Phys. Lett. **68**, 2699–2701 (1996)
64. Wu, J., Becerril, H.A., Bao, Z., Liu, Z., Chen, Y., Peumans, P.: Appl. Phys. Lett. **92**, 263302 (2008)
65. Yin, Z., Sun, S., Salim, T., Wu, S., Huang, X., He, Q., Lam, Y.M., Zhang, H.: ACS Nano **4**, 5263–5268 (2010)

66. Murray, I.P., Lou, S.J., Cote, L.J., Loser, S., Kadleck, C.J., Xu, T., Szarko, J.M., Rolczynski, B.S., Johns, J.E., Huang, J.: J. Phys. Chem. Lett. **2**, 3006–3012 (2011)
67. Stratakis, E., Savva, K., Konios, D., Petridis, C., Kymakis, E.: Nanoscale **6**, 6925–6931 (2014)
68. Lu, Y., Jiang, Y., Wei, W., Wu, H., Liu, M., Niu, L., Chen, W.: J. Mater. Chem. **22**, 2929–2934 (2012)
69. Lee, B.R., Kim, J.W., Kang, D., Lee, D.W., Ko, S.J., Lee, H.J., Lee, C.L., Kim, J.Y., Shin, H.S., Song, M.H.: ACS Nano **6**, 2984–2991 (2012)
70. Wang, D.Y., Wang, I.S., Huang, I.S., Yeh, Y.C., Li, S.S., Tu, K.H., Chen, C.C., Chen, C.-W.: J. Phys. Chem. C **116**, 10181–10185 (2012)
71. Jiang, X.C., Li, Y.Q., Deng, Y.H., Zhuo, Q.Q., Lee, S.T., Tang, J.X.: Appl. Phys. Lett. **103**, 073305 (2013)
72. Bi, Y.G., Feng, J., Li, Y.F., Zhang, Y.L., Liu, Y.S., Chen, L., Liu, Y.F., Guo, L., Wei, S., Sun, H.-B.: ACS Photon. **1**, 690–695 (2014)

Chapter 5
Application of GO in Energy Conversion and Storage

Abstract The increasing depletion of fossil fuel inspires the demand for renewable energy and energy-efficient devices. GO-based materials emerge in applications of energy storage and conversion with superior advantages. Especially, the composition of GO with specified materials not only retains the inherent characteristics of GO but also induces various characters for improving the performance in energy conversion and storage. Due to the large surface area and ample oxygen containing functional groups, GO can bind many active materials or catalysts for hydrogen storage and generation. Moreover, the functional groups enable GO to further couple with other species and thus form various porous or hierarchical architectures as electrodes, electrolyte or current collector in the lithium batteries and supercapacitors.

5.1 Hydrogen Production and Storage

5.1.1 Photocatalytic Water Splitting

5.1.1.1 GO as a Photocatalyst

As already discussed in Chap. 3, the electronic properties of GO rely on the specific composition. GO exhibits p-type conductivity because of oxygen's high electronegativity compared to carbon. Likewise, n-type conductivity appears when graphene covalently bonds to electron donating nitrogen-containing functional groups [1]. As oxygen bonds to graphene, the valence band originates from the O 2p orbital rather than the π orbital of graphene, leading to a larger band gap for a high oxygen coverage rate of GO and thus an evolution from semimetal to insulator. Meanwhile, the conduction band of GO is mainly contributed by the anti-bonding π^* orbital, which has a higher energy level than that needed for H_2 generation [2]. Therefore, GO with an appropriately functionalization might be a promising photocatalyst [1–7].

© The Author(s) 2015   79
J. Zhao et al., *Graphene Oxide: Physics and Applications*,
SpringerBriefs in Physics, DOI 10.1007/978-3-662-44829-8_5

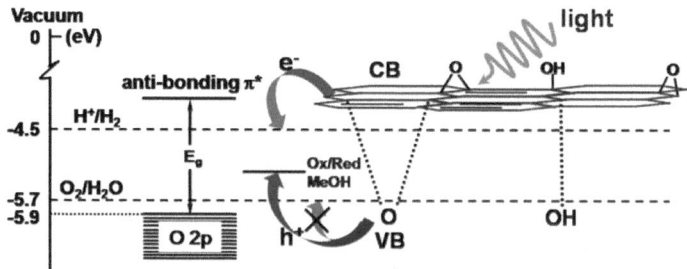

Fig. 5.1 Schematic energy-level diagram of GO relative to the levels for H_2 and O_2 generation from water. Reprinted with permission from Ref. [2]. Copyright (2010) WILEY-VCH Verlag GmbH & Co. KGaA, Weinheim

Yeh et al. [2] demonstrated photocatalytic H_2 evolution activity of GO with a band gap of 2.4–4.3 eV. Electrochemical analysis along with the Mott-Schottky equation illustrates that GO exhibits stable H_2 generation from an aqueous methanol solution or pure water under mercury light irradiation, even in absence of the Pt cocatalyst (Fig. 5.1). Afterwards, they investigated the photocatalytic activity of GO at various oxidation levels and established an inverse relationship between the amount of H_2 evolution and the population of the oxygen-containing groups on the GO sheets [5]. They concluded that the GO sheets with higher oxidation degree have a larger band gap and limited absorption of light, thus exhibiting a lower photocatalytic activity than the GO sheets with lower oxidation degree. In addition, Matsumoto et al. [7] reported the photoreactions to generate H_2 from an aqueous suspension of GO nanosheets under UV irradiation and also found that the GO with an appropriate reduction level can serve as a photocatalyst for H_2 production.

Considering the p-type conductivity which hinders hole transfer for water oxidation and suppresses O_2 evolution, Yeh et al. [1] introduced amino and amide groups on the GO surface and demonstrated the ammonia-modified GO exhibits n-type conductivity and is able to catalyze the H_2 and O_2 evolution simultaneously. To further investigate the overall photocatalyst of GO, Yeh et al. [6] synthesized nitrogen-doped graphene oxide quantum dots (NGO-QDs) as the catalyst to fulfill the evolution of H_2 and O_2 at a molar ratio of approximately 2:1. This can be explained that p-n diodes configuration of NGO-QDs results in an internal Z-Scheme charge transfer for effective reaction at the QD interface. For this reason, visible light irradiation on the NGO-QDs leads to simultaneous H_2 and O_2 evolution from the pure water.

To clarify the optimal composition of GO for high photocatalytic activity, Zhao's group [8] has investigated the key electronic properties of GO that are responsible for photocatalytic water splitting using DFT calculations. The effects of epoxy and hydroxyl functionalization on the work function, band gap, positions of conduction band minimum (CBM) and valence band maximum (VBM), and optical absorption spectra are discussed. Their GO structures are

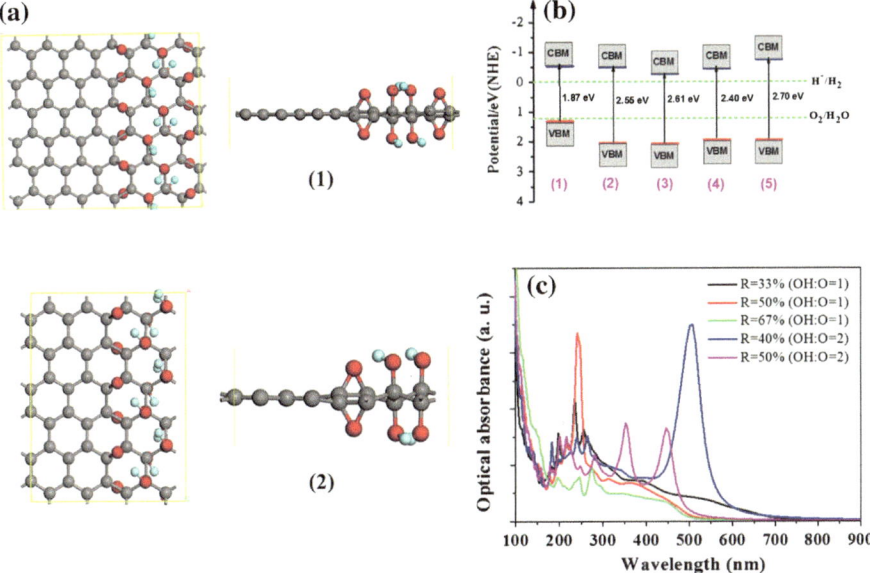

Fig. 5.2 a Representatively structural models of GOs with coverage rate of R = 50 % after geometry relaxation: *1* top view (*left*) and side view (*right*) of GO with OH:O = 1:1; *2* top view (*left*) and side view (*right*) of GO with OH:O = 2:1. **b** Site levels of VBM and CBM for OH:O = 1 and OH:O = 2 graphene oxide with different coverage rate R: (*1*), (*2*), (*3*), (*4*), and (*5*) represented $C_{36}O_8H_4$, $C_{24}O_8H_4$ and $C_{24}O_{12}H_6$ with OH:O = 1, $C_{20}O_6H_4$ and $C_{16}O_6H_4$ with OH:O = 2, respectively. The *dot lines* are standard water redox potentials. The reference potential is the vacuum level. **c** Optical absorption curves for GO with OH:O = 1:1 and OH:O = 2:1 under different coverage rates. Reprinted with permission from Ref. [8]. Copyright (2012) Elsevier Inc.

based on the stable structural models with different OH:O ratios and coverage proposed by the same group [9, 10], which have been illustrated in Chap. 3. As shown in the Fig. 5.2a, the epoxy and hydroxyl groups aggregate along the arm-chair direction and form stable one-dimensional (1-D) chain configurations on the basal plane.

With increasing coverage, the band gap of epoxy and/or hydroxyl functional-ized graphene sheets is continuously tunable from metallic to insulating, and the work function of GO also increases. By varying the coverage and relative ratio of the epoxy and hydroxyl groups, both the band gap and work function of certain GO systems can meet the requirements of photocatalyst. The redox energy levels of GOs with respect to the water oxidation/reduction potential levels and the simulated optical absorption spectra are displayed in Fig. 5.2b, c, respectively. The electronic structures of GO materials with 40–50 % (or 33–67 %) coverage and OH:O ratio of 2:1 (or 1:1) are suitable for both reduction and oxidation reac-tions of water splitting. Among them, the GO composition with 50 % coverage and OH:O = 1:1 is most promising for visible light driven photocatalyst [8].

Besides, metal deposition can enhance the photocatalytic activity of GO. For example, Agegnehu et al. [11] found that the H_2 generation rate of Ni/GO composite from aqueous methanol solution under UV-visible light illumination is enhanced by approximate four to seven times compared to that of the bare GO.

5.1.1.2 Binary and Ternary Composites by GO and Semiconductor

In the past years, TiO_2 is the most widely used photocatalyst because of its high activity, chemical inertness, low cost and non-toxicity. However, the high recombination rate between the photo-generated electrons and holes, the fast backward reactions between hydrogen and oxygen in water restrict the hydrogen production activity of the TiO_2 photocatalysts. In fact, such unbeneficial factors almost influence all of other visible light active semiconductors [12]. Noble metal (Pt, Pd, Ru, Ag and etc.) loaded on semiconductor surfaces as cocatalyst is a feasible strategy to significantly enhance the hydrogen production for photocatalytic water splitting [13, 14]. However, noble metals are rare, expensive and harmful to the environment.

To tackle the above problems, many efforts have been devoted to designing GO and semiconductor hybrid composites to replace noble metal, taking the advantages of superior electron mobility, large specific surface area and high work function of GO. To date, large numbers of efficient GO-based composites for photocatalytic water splitting have been synthesized, including GO-semiconductor binary systems and more complicated ternary composites [12, 15–17]. For instance, Li et al. [18] found that GO-CdS nanocomposites at GO content of 1.0 wt% and Pt 0.5 wt% reach a high H_2 production rate of 1.12 mol h^{-1} under visible light irradiation, corresponding to an apparent quantum efficiency of 22.5 % at wavelength of 420 nm, which is enhanced by 4.84 times in comparison with that of the pure Pt-CdS. Another example of high quantum efficiency is GO-$Zn_{0.8}Cd_{0.2}S$ system [19], whose H_2 production rate and quantum efficiency are 1,824 μ mol h^{-1} g^{-1} and 23.4 % at 420 nm, respectively, much better than the Pt-$Zn_{0.8}Cd_{0.2}S$ photocatalyst (Fig. 5.3). In addition, binary composites of GO/TiO_2 [20, 21], GO/AgBr [22], GO/g-C_3N_4 [23], GO/3C-SiC [24], GO/Cu_2O [25], GO/$Sr_2Ta_2O_{7-x}$ [26] have also been investigated; all of them showing higher photocatalytic H_2 evolution than the bare semiconductors.

Hou et al. [27] further explored other ternary GO composites and developed a simple hydrothermal-assisted ion-exchange route to obtain stable and high H_2 evolution efficiency using GO nanosheets decorated with CdS sensitized TaON core-shell composites (GO-CdS@TaON). The TaON core-shell composites containing 1 wt% CdS nanocrystals shows a high rate of hydrogen production at 306 μ mol h^{-1} with an apparent quantum efficiency of 15 % under 420 nm monochromatic light. Other ternary GO-based composites with high photocatalytic behavior include CdS/Al_2O_3/GO [28], CdS/ZnO/GO [28], and TiO_2/MoS_2/GO [29].

The enhancement of H_2 production rate is mainly ascribed to the role of GO as an electron acceptor and transporter to separate photogenerated electron and

Fig. 5.3 a Schematic illustration for the charge transfer and separation in the GO-Zn$_{0.8}$Cd$_{0.2}$S system. **b** The mechanism for photocatalytic H$_2$ production under simulated solar irradiation. Reprinted with permission from Ref. [19]. Copyright (2012) American Chemical Society

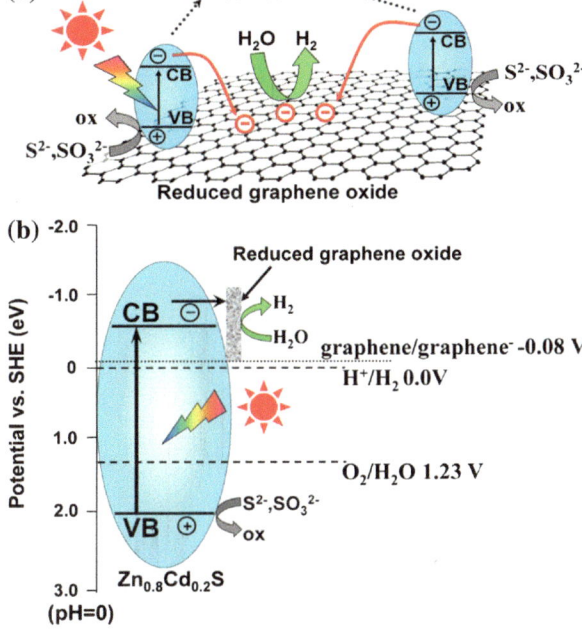

hole pairs. Experimentally, the charge transfer dynamics of GO-based composites (e.g., GO/CdS) were analyzed by transient absorption spectroscopy and transient photovoltage technique [18]. It was demonstrated that the electron and hole are efficiently separated by transferring photoinduced electrons from CdS to GO, and the recombination of electron and hole pairs in the excited semiconductor material is suppressed simultaneously. Using first-principles calculations, Dong et al. [30] investigated the interfacial electron-hole separation mechanism of GO-CdS nano-materials and found that the excited electrons in CdS are injected into GO and transport along graphene layer through π* orbital under visible light irradiation to achieve electron and hole separation, in agreement with the experiment [18].

5.1.1.3 GO as a Mediator: Z Scheme Photocatalyst

Alternatively, GO can be used as a solid-state electron mediator for water splitting in the Z-scheme photocatalysis system [31], which can overcome the problems of electronic recombination and transportability. In brief, the Z-scheme system is a two-proton system, consisting of a H$_2$ evolving photocatalyst, an O$_2$ evolving photocatalyst, and an electron mediator. Generally, the electron transfer between two distinct photocatalysts in the Z-scheme photocatalysis system is the most important factor for producing H$_2$ and O$_2$ from water splitting.

Amal's group [16] constructed a model system with GO as solid electron mediator, BiVO$_4$ as O$_2$ photocatalyst, and Ru/SrTiO$_3$:Rh as H$_2$ photocatalyst,

Fig. 5.4 **a** Schematic
image of a suspension of
Ru/SrTiO₃ and GO/BiVO₄
in water. **b** Mechanism of
water splitting in a Z-scheme
photocatalysis system
consisting of Ru/SrTiO₃:Rh
and GO/BiVO₄ under visible-
light irradiation. Reprinted
with permission from Ref.
[16]. Copyright (2011)
American Chemical Society

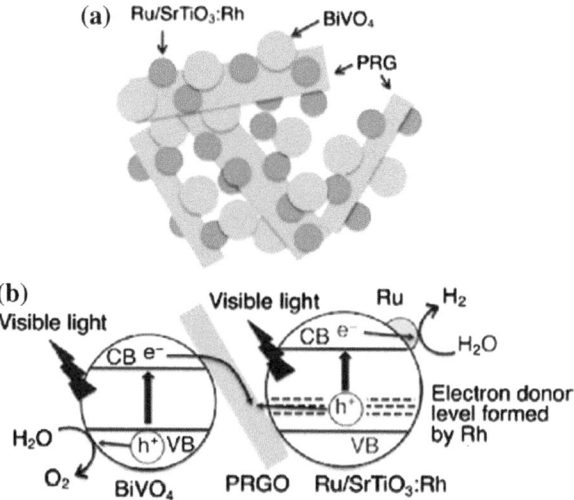

respectively. Such system shows higher H_2 production rate than GO/BiVO₄ and Ru/SrTiO₃:Rh. The mechanism of electron transfer process can be described as GO acting as an electron mediator to transfer electrons from the conduction band of BiVO₄ to the Ru/SrTiO₃:Rh. The electrons in Ru/SrTiO₃:Rh reduce water to H_2 on the Ru cocatalyst, while the holes in BiVO₄ simultaneously oxidize water to O_2, accomplishing a complete water splitting cycle, as displayed in Fig. 5.4.

As discussed above, GO and GO-based composites not only show the ability in separating the photogenerated electron-hole pair, but also exhibit the capacity for photocatalytic H_2 evolution by itself. However, research on GO-based materials for H_2 generation from light driven water splitting is still at its primary stage and requires further attentions. First, explanation of photocatalytic activity by GO content in these composites is still controversial. Second, the mechanism of photocatalytic reaction is partly unclear. Finally, theoretical calculations are highly desirable to provide some insightful guidelines.

5.1.2 Physical Hydrogen Storage

5.1.2.1 GO and Metal-Decorated GO for Hydrogen Storage

In previous report, the interaction energy between a H_2 molecule and pristine graphene is only 1.2 kJ/mol [32], which is too weak to hold the H_2 molecules stably at room temperature, and therefore leading to very low storage capacity of hydrogen. To improve it, Kim et al. proposed a feasible way to modulate the interlayer distance of multilayered graphene oxide (GO) by thermal annealing [33]. They found that the hydrogen storage capacity depends on the interlayer distance of GO

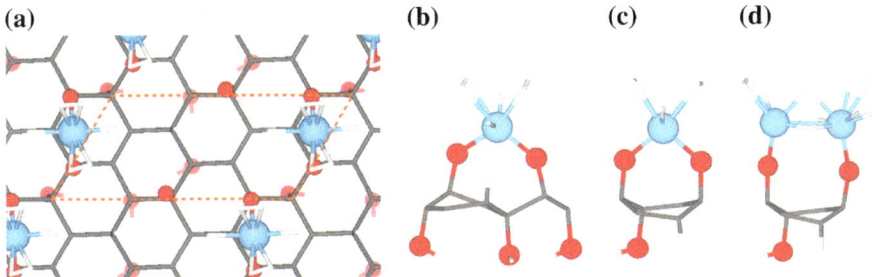

Fig. 5.5 Ti-decorated GO with multiple H_2 adsorption. **a** *Top view* H_2 binding to Ti@Z2. **b–d** *Side views* H_2 binding to Ti@Z3, A3, and Ti_2@A3, respectively. Reprinted with permission from Ref. [10]. Copyright (2009) American Chemical Society

and an optimal distance of 6.3 Å results in a maximum hydrogen storage capacity of 4.8 (0.5) wt% at 77 K (298 K) and 9.0 MPa pressure in three GO-amine composites. Furthermore, they investigated the pore size dependence of thermally RGO for hydrogen storage and demonstrated that 6.7 Å is the optimum pore size, leading to a maximum capacity of 5.0 wt% at 77 K [34].

Moreover, the functional groups on GO surface offer feasible ways of further decoration with other compounds to improve the H_2 binding energy. In principle, metal (especially transition metal) decoration on carbon sorbents is a promising route to enhance the H_2 binding [35], except for the huge obstacle of metal aggregation [36]. To achieve strong interaction between sorbents and metal atoms, Wang et al. [10] proposed to use GO to support and disperse Ti atoms without clustering. Their first-principles computations show that the Ti atoms can be stably anchored by the hydroxyl groups on GO surface and simultaneously retain sufficient activity to adsorb H_2 molecules. As shown in Fig. 5.5, each Ti is able to bind multiple H_2 with the desired binding energies (14–41 kJ/mol), corresponding to a theoretical gravimetric (volumetric) density of 4.9 wt% (64 g/L). Soon later, the hydrogen adsorption properties of Pd-doped GO were investigated in experiment [37], which demonstrated that the Pd decoration can extremely enhance the hydrogen storage ability in comparison with that of pristine GO. To further simplify the synthesis process of GO/transition metal oxide (TMO) composites, Kim et al. put forward the GO wrapped TMO composite materials without any additional agents and improved the hydrogen storage capacity up to 1.36 wt% for GO/V_2O_5 (1.26 wt% for GO/TiO_2) compared to the 0.16 wt% for V_2O_5 alone (0.58 wt% for TiO_2) [38].

Apart from the transition metals, Chen et al. [39] found that Mg doping can reduce the hydroxyl group from GO surface no matter whether the hydroxyl exhibits an acidity or alkalinity. After that, the remain Mg atoms can be strongly bound on GO surface in the form of –(C–O)$_x$–Mg (x = 1 or 2) without clustering. H_2 molecules can be polarized by the strong electric field jointly produced by the anchored Mg and O on GO surface, and therefore leading to a favorable binding of four H_2 per Mg with an average binding energy of 28 kJ/mol. Accordingly, the hydrogen storage capacity reaches 5.6 wt% at a temperature of 200 K without any pressure.

5.1.2.2 GO-Based 3-D Frameworks for Hydrogen Storage

Besides metal atoms, GO can be also composited with other compounds like mul-
tiwalled carbon nanotubes (MWCNTs), metal organic frameworks (MOFs) or
other species which can react with the functional groups for hydrogen storage.
By composition, GO-based three-dimensional (3-D) porous materials can either
achieve higher interlayer space or increase the effective sites for H_2 adsorption.
For instance, Froudakis et al. [40] proposed a strategy of combining GO and CNT
and designed a new class of pillared graphene oxide, which increases the surface
area to accommodate more H_2 in-between the layers. To improve the H_2 binding
energy, they replaced the OH groups with the O–Li groups which effectively prevent the Li aggregation. Their computations show that Li decorated pillared GO
with pore dimensions of d = 23 Å and an O/C ratio of 1/8 can achieve a gravimetric capacity greater than 10 wt% and a volumetric capacity of 55 g/L at 77 K
and 100 bar. Later, Aboutalebi et al. [41] experimentally demonstrated that the
processability of GO dispersions can be further exploited to fabricate self-aligned
GO-MWCNT hybrid frameworks (see Fig. 5.6). Intercalation of MWCNT as 1-D

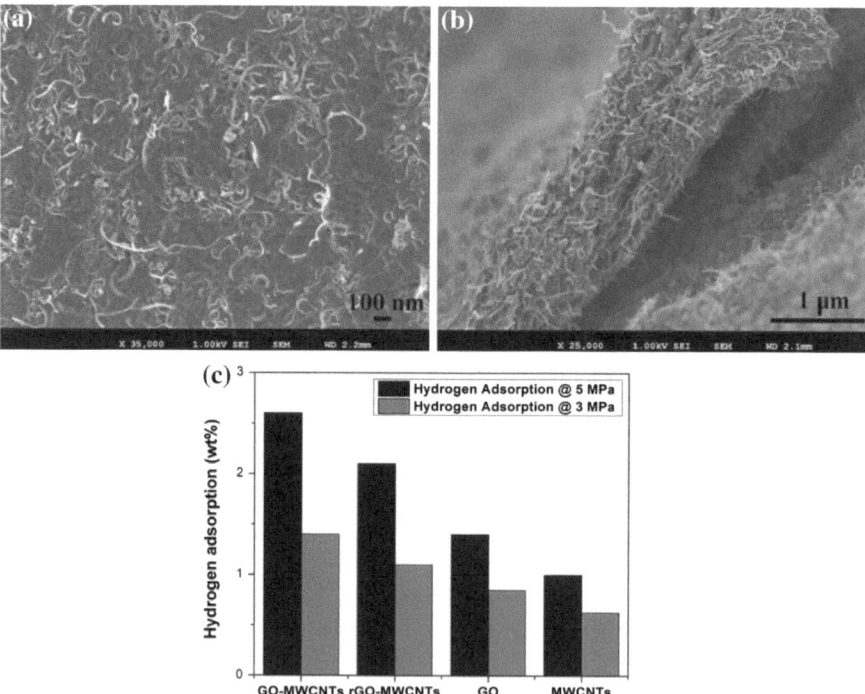

Fig. 5.6 **a** MWCNTs distributed on GO layer; **b** layer-by-layer assembled GO platelets decorated by MWCNTs. **c** Comparative hydrogen adsorption of GO, MWCNTs, GOMWCNTs and
rGO-MWCNTs at different hydrogen pressures. Reprinted with permission from Ref. [41].
Copyright (2012) WILEY-VCH Verlag GmbH & Co. KGaA, Weinheim

spacer within GO-MWCNT provides proper interlayer distance, leading to an appreciable hydrogen uptake of 2.6 wt% at room temperature.

By assembling with reactive agents through the functional group on surface, GO can form a new class of 3-D pillared porous materials with tunable porosity, accessible surface area and versatile electronic properties. Prud'homme et al. [42] synthesized the pillared GO (intercalated with diaminoalkanes) via tailored interlayer spacing. Later, Yildirim and coworkers [43] intercalated the boronic acid (which can react with the hydroxyl groups) into GO layers and successfully synthesized a class of graphene oxide frameworks (GOFs) with 3-D porosity, as displayed in Fig. 5.7. Grand canonical Monte Carlo (GCMC) simulation of adsorption isotherms for several representative GOF structures predicted the GOF-32 with one linker per 32 graphene carbon atoms possesses an H_2 adsorption capacity of 6.1 wt% at 77 K and 1 bar; but the experimentally measured H_2 adsorption capacity is 1.0 wt% at 77 K and 1 bar. Yildirim and coworkers [44] further synthesized a range of porous GOFs with strong boronate-ester bonds between GO layers. By adopting various linear boronic acid pillaring units, they were able to tune the interlayer spacing between graphene planes to an optimum amount for H_2 adsorption on both surfaces. The GOFs exhibit high isosteric heat of adsorption and hydrogen adsorption capacity (twice of typical porous carbon material and comparable to MOFs). Based on the reported porous GOFs, Chan et al. [45] investigated the hydrogen storage properties of GOFs (GOF-120, GOF-66, GOF-28 and GOF-6) with a mathematic model and found that the GOF-28 has

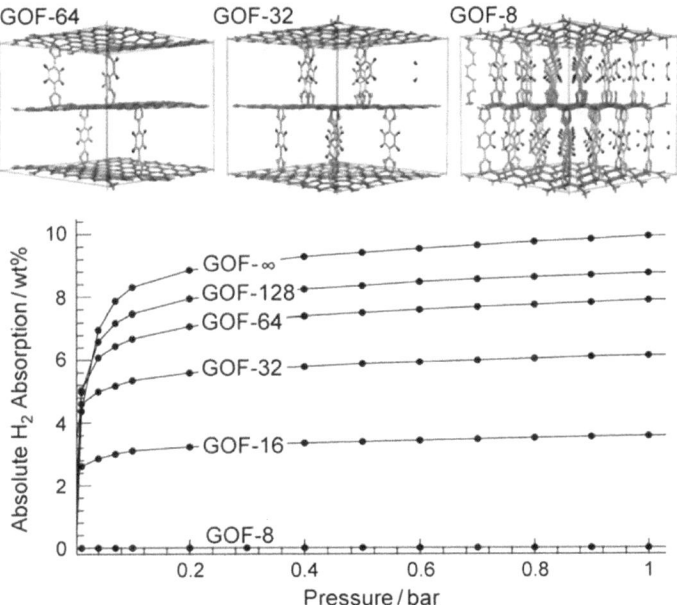

Fig. 5.7 GCMC simulations for ideal GOF-n structures with n graphene carbon atoms per linker. The structures of three examples with n = 64, 32, 8 are shown. Reprinted with permission from Ref. [43]. Copyright (2010) WILEY-VCH Verlag GmbH & Co. KGaA, Weinheim

highest hydrogen uptake of 6.33 wt%. The outstanding hydrogen storage properties are attributed to the porous spaces; most importantly the enhanced hydrogen adsorption is caused by the benzenediboronic acid pillars between graphene sheets.

To search the best interlayer distance of 3-D pillared GO for hydrogen adsorption, Kim et al. [46] tuned the intercalation via three kinds of diaminoalkanes. They found an optimum GO interlayer distance for a maximum H_2 uptake at 6.3 Å, similar to the predicted distance from thermally modulated GO [33]. Using the RGO, Kumar et al. [47] recently synthesized two porous graphene frameworks (PGFs) via a C–C coupling reaction between RGO and iodobenzene. Both the PGFs exhibit tunable porosity and surface area with hydrogen storage capacity (at 77 K) of 1.2 wt% at 1 atm and 1.9 wt% at 20 atm, respectively.

Except for the intercalation into GO layers to enlarge the space for accommodating hydrogen molecules, intercalation of MOFs into GO layers also increases the effective sites to attract H_2. Even with the high porosities, most MOFs cannot restrain H_2 molecules due to the large size pores. After intercalation, the GO/MOFs composition creates suitable pores size for H_2 adsorption. Petit and Bandosz [48] firstly proposed the concept of GO/MOFs composites and synthesized a class of these composites with various ratios (see Fig. 5.8). Afterwards, they synthesized

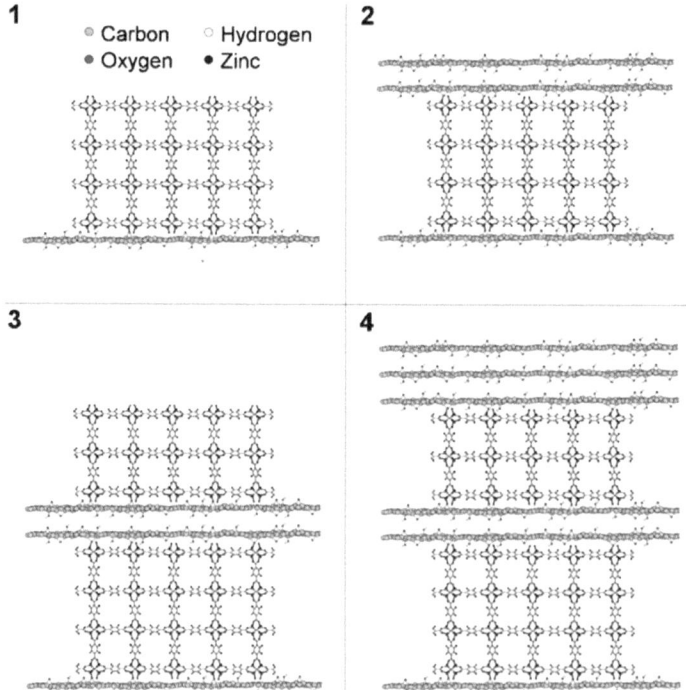

Fig. 5.8 Schematic view of the steps (1–4) of the nanocomposite (MOF-5–GO1) formation. Reprinted with permission from Ref. [48]. Copyright (2009) WILEY-VCH Verlag GmbH & Co. KGaA, Weinheim

a series of copper-based GO/MOFs composites and observed the enhancement of hydrogen uptake, therefore confirming the formation of new small pores [49].

Later, Liu et al. [50] also synthesized the Cu based MOF-GO composites for gas storage. They found that nanosized Cu-BTC (Copper-benzene-1, 3, 5-tri-carboxylate) is formed and well dispersed by the incorporation of GO, which shows great improvement of hydrogen storage capacity compared to the pristine Cu-BTC (from 2.81 wt% of Cu-BTC to 3.58 wt% of CG-9 at 77 K and 42 atm). Additionally, a new class of metallomacrocycle–graphene frameworks (MGFs) via tunable porosity from microporous to hierarchical micro- and mesoporous, has been reported by Kim et al. [51]. Their fabrication route enables alternative stacking of GO and Ni (II/III) metallomacrocycles in a layer-by-layer manner and allows the usually unstable square planar Ni (III) species to be stabilized in the solid state. The hydrogen uptake of the composites is affected by the structural modifications under different synthetic conditions, with highest hydrogen uptake of 1.54 wt% at 77 K and 1 bar.

Recently, Zhou et al. [52] synthesized the Pt@GO/HKUST-1 hybrid composites. The Pt nanoparticles (NPs) are well-dispersive and anchored tightly into composites, so that hydrogen molecules can be chemisorbed and dissociate on the Pt surface, migrate to the GO surface, and further diffuse into HKUST-1 with high porosity and new pores. A high hydrogen uptake of 0.77 wt% at 80 bar and 298 K, which is nearly twofold enhancement with regard to HKUST-1, can be reasonably ascribed to the spillover mechanism and the high porosity of the secondary receptor HKUST-1.

5.1.3   Chemical Hydrogen Storage

To release the stored hydrogen in hydrates, two major ways, i.e., thermo-dehydrogenation and hydrolytic dehydehydrogenation in solvent have been utilized. However, both of them show high temperature requirement or slow reaction kinetics without any auxiliary. Transition metal NPs, with intriguing structural, electronic and magnetic properties, therefore serve as promising catalysts for hydrogen extraction. To optimize the catalysis performance, the size and morphology of metal NPs have to be well-controlled. However, reduction of particle size is concomitant with the increase of surface energy, which usually leads to serious aggregation of the small particles in the absence of protective agents and diminished catalysis performance in the practical applications [53]. To avoid aggregation, RGO emerges as a suitable support to disperse and stabilize the NPs for chemical hydrogen storage owing to its large surface area and excellent chemical stability.

Xi et al. [54] demonstrated a wet chemistry synthesis of RGO/Pd nanocomposite [55] for the hydrolytic dehydrogenation of AB. The small-sized Pd NPs are firmly attached and well dispersed on the RGO sheets without any surfactants. These RGO/Pd nanocomposite exhibit enhanced catalytic activity in hydrogen generation of AB hydrolysis. The performance of hydrolysis completion time

(12.5 min) and activation energy (51 ± 1 kJ/mol) of the RGO/Pd nanocomposite is comparable to the best Pd-based catalyst reported. Another route of synthesizing RGO/Pd nanocomposite for AB hydrolysis was reported by Metin and coworkers [56]. They also demonstrated high activity of RGO/Pd nanocomposite in the hydrolytic dehydrogenation of AB. Even more, their RGO/Pd nanocomposite shows slightly lower activation energy of 40 ± 2 kJ/mol and higher turnover frequency (TOF) than Xi's results [58].

In addition to Pd, composites of GO with other noble metal catalysts have been also investigated [57–61]. A facile route to synthesis Ru/RGO NPs via the methylamine borane (MeAB) as reducing agent was firstly reported by Cheng et al. [57]. Their experiments provide the ever reported lowest activation energy of 11.7 kJ/mol for the catalytic hydrolytic dehydrogenation of AB, demonstrating the superior catalytic activity of Ru/RGO NPs. To further improve the catalytic activity, they synthesized the RGO supported core-shell NPs of Ag@M (M = Co, Ni, Fe) [58] and Ru@Ni [59] for the hydrolysis of AB and MeAB. Among the three Ag@M NPs, Ag@Co/RGO exhibits the highest catalytic activity, followed by Ag@Ni. The high TOF value of 102.4 and low activation energy of 20.3 kJ/mol are obtained for Ag@Co/RGO catalyzed AB hydrolysis. Meanwhile, Ru@Ni/RGO NPs facilitate the hydrolysis of AB (MeAB) with the TOF value of 340, which is among the highest values reported on Ru-based NPs so far; but the activation energy of ~37.1 kJ/mol is slightly larger than the case of Ag@Co/RGO. Furthermore, the core-shell NPs of Ag@Co/RGO and Ru@Ni/RGO show good recyclability and magnetically reusability for the hydrolytic dehydrogenation of AB and MeAB for reusing of the catalysts.

Zhang et al. also demonstrated the catalytic activity of another noble metal Pt–CeO$_2$ NPs supported by RGO [60]. The hybrid metal oxide and RGO exhibit synergistic effect to stabilize the active centers and increase their catalytic activities. However, the TOF value of the hybrid Pt–CeO$_2$/RGO NPs is only 48.0, much lower than the catalysts with core-shell structure [58–61].

Since noble metals have limited resource in nature, it is desirable to utilize the alternative non-noble metal composites for catalytic applications [62–66]. Lu et al. proposed to use the GO supported Fe–Ni NPs with the auxiliary of polyethyleneimine for effectively suppress the metal aggregation [62]. The content of PEI attached on GO significantly affect the morphology and size of resulting Fe–Ni NPs, and therefore modulate the catalytic activity. Compared to the Fe–Ni NPs directly deposited on GO, the NPs on PEI-decorated GO has a dehydrogenation rate of 982 ml min^{-1} g^{-1} at 293 K for the hydrolysis of AB. Very recently, RGO supported Cu NPs were reported as catalyst for the hydrolytic dehydrogenation of AB, which exhibits the highest TOF value of 3.61 among all of the Cu nanocatalysts ever reported for this reaction [63].

For the H$_2$ release through thermo-dehydrogenation, Tang et al. [65] realized a recyclable dehydrogenation of AB within a GO-based hybrid nanostructure. They showed that the hydroxyl groups on the GO surface act as a proton donor to react with AB and yield H$_3$NBH$_2^+$ cation, which is the key to facilitating AB dehydrogenation (see Fig. 5.9). The combined modification strategy of acid activation and

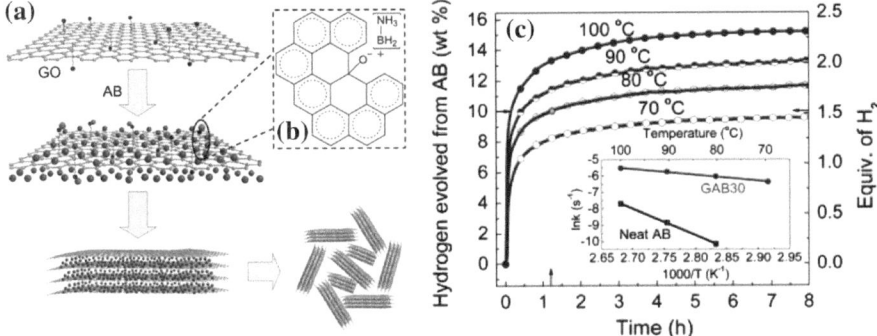

Fig. 5.9 a Schematic representation of the mechanism of the formation of GO-AB hybrid nano-structure. **b** Detailed illustration of interaction between AB cation and negatively charged oxygen of GO. **c** Time-extended results of hydrogen released from AB in GAB30 (30 wt% AB loaded on GO). Reprinted with permission from Ref. [65]. Copyright (2012) American Chemical Society

nanoconfinement by GO allows AB to release more than 2 equiv of pure H_2 at temperatures below 100 °C.

Recently, Li et al. [66] proposed to combine lithium amidoboranes (LiAB) with GO to form a hybrid complex since the hydroxyl groups on the GO surface may interact with LiAB via one molar equivalent of H_2 released. Compared to the pure LiAB, the hybrid GO–LiAB complex shows greater dehydrogenation performance for chemical hydrogen storage, according to their calculations of minimum energy pathway for the dehydrogenation process. Most strikingly, using the dehydrogenated products of GO-LiAB complex for physisorption of H_2 molecules, 5 wt% of hydrogen can be stored. In such a way, chemisorption and physisorption can be combined for superior hydrogen storage.

5.2 Lithium Batteries

5.2.1 Lithium Ion Batteries

Lithium-ion batteries (LIBs) are one of the most popular rechargeable batteries for critical applications such as electric vehicles, electronic devices, locomotives, and aerospace. However, the theoretical capacity limits with the conventional electrode materials impede its further applications. It is imperative to search novel LIB materials with high reversible capacity, long cycle life, and low cost. In this regard, elaborately designed GO/RGO based composites exhibit superior performance in both anode and cathode materials [67–71].

Direct combination of GO with the commercial graphite as binder-free anode material in LIBs improves the reversible capacity to 690 mAh g^{-1} at the rate of 0.5 C (1 C = 372 mA g^{-1}), with excellent cycle performance and rate capability simultaneously [67]. The original graphite maintains the conductivity of the system, while the GO mainly contributes to the enhanced capacity as the lithium accommodator. The conductive RGO itself can replace the conventional graphite in anode. More importantly, the residual functional groups provide wide range of the porous structure formation, which benefits the electrochemical performance of LIBs [69–71]. The photothermal reduced GO leads to open-pore structures with micrometer-scale pores, cracks, and intersheet voids (see Fig. 5.10), which provide more space available for lithium accommodation and also facilitate the intercalation kinetics [71]. Moreover, at charge/discharge rates of ~40 C, the photoreduced GO anodes exhibit a steady capacity of ~156 mAh g_{anode}^{-1} (power density of ~10 kW/kg$_{anode}$) continuously over 1,000 charge/discharge cycles, showing outstanding stability and cycling ability.

Transition metal oxides also exhibit better Li^+ insertion/extraction performance than the conventional graphite. But the metal oxides (MOs) electrodes suffer from rapid degradation in capacity due to the pulverization process. To improve the

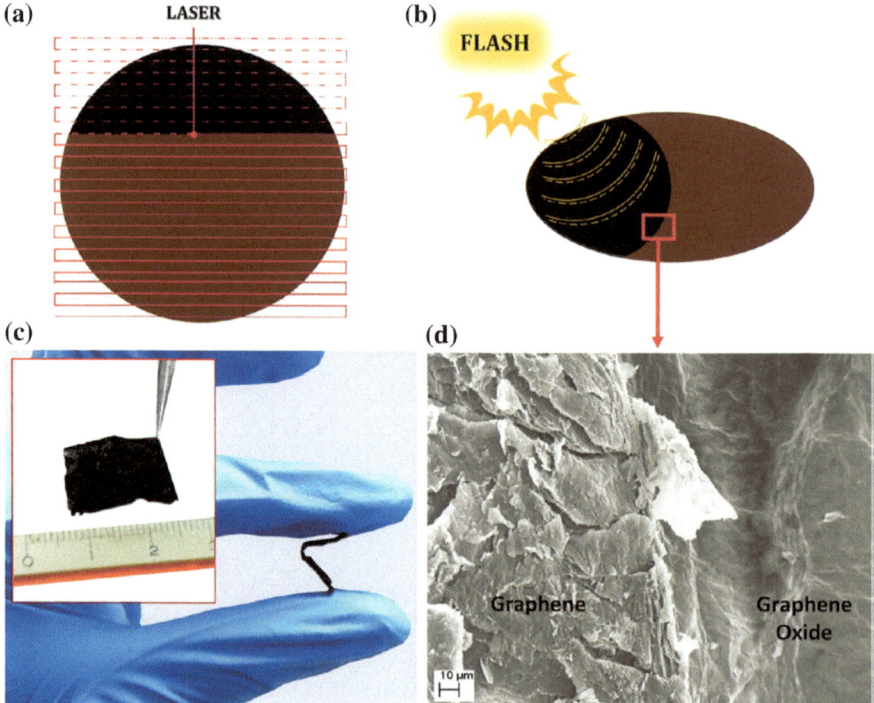

Fig. 5.10 Photothermal reduction of GO. **a** Schematic showing the laser reduction system with the raster scan path. **b** Schematic showing the flash reduction of GO whereby a flash from a digital camera reduces graphene oxide. **c** Photograph of the flash-reduced graphene, displaying its structural integrity. **d** SEM image showing two characteristic regions of flash-reduced graphene and graphene oxide. For this, a part of the GO sample was intentionally shielded from the photoflash in order to illustrate the contrast in porosity between the two regions. Reprinted with permission from Ref. [71]. Copyright (2012) American Chemical Society

cyclability of electrode, composites of MOs (e.g., Mn_3O_4, Co_3O_4, SnO_2, FeO_x) with RGO have been intensively investigated as anode materials in LIBs [72–76]. In addition to acting as the electronic conductor, the RGO also affect the configuration and stability of the anode materials and result in excellent electrochemical performance. As shown in Fig. 5.11, Wang et al. [72] synthesized a hybrid material by selective growth of Mn_3O_4 NPs on RGO sheets, in which the Mn_3O_4 nanoparticles are wired up to a current collector through the underlying conducting graphene network to maintain the conductivity. The intimate interactions between the RGO substrates and the Mn_3O_4 NPs render a high specific capacity up to ~900 mAh g^{-1} (near the theoretical capacity) as well as good rate capability and cycling stability. Despite the high performance, synthesis of the Mn_3O_4/RGO composites requires harsh solvents and multiple-steps, which hinder the large-scale preparation.

Since that, more effective strategies to synthesize the MO/GO hybrids were developed [85–100]. Synthesized by a one-step facile and environment-friendly way, Fe_2O_3/CoO-RGO composites used as anodes for LIBs show high capacities and excellent charge-discharge cycling stability in the voltage window between 0.01 and 3.0 V [73]. The Fe_2O_3 NPs/RGO hybrids possess specific capacity of 881 mAh g^{-1} in the 90th cycle at a discharge current density of 302 mA g^{-1}. The efficient diffusion of Li ions and high specific capacities are attributed

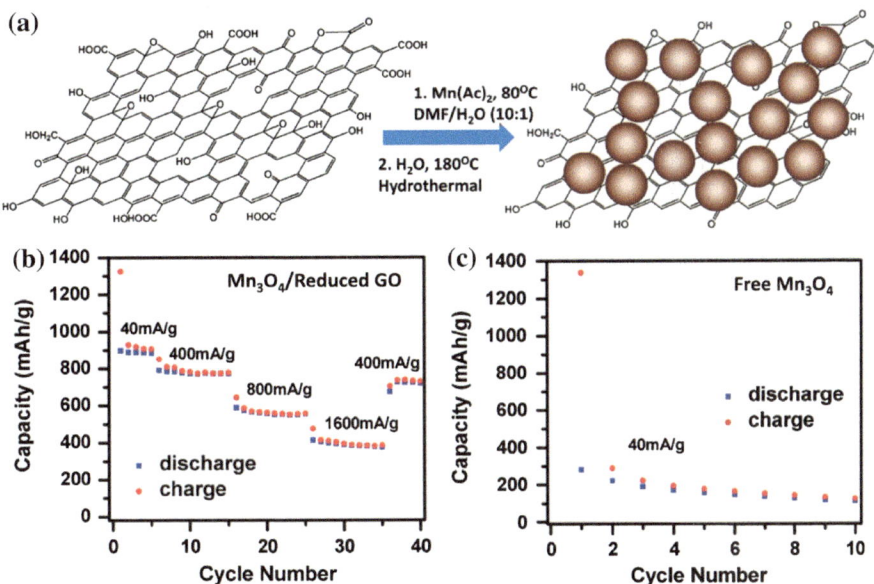

Fig. 5.11 Mn_3O_4 nanoparticles grown on GO. **a** Schematic two-step synthesis of Mn_3O_4/RGO. **b** Capacity retention of Mn_3O_4/RGO at various current densities. **c** Capacity retention of free Mn_3O_4 nanoparticles without graphene at a current density of 40 mA g^{-1}. Reprinted with permission from Ref. [72]. Copyright (2010) American Chemical Society

to the formation of small crystalline grains, facilitated by GO or RGO as the heterogeneous nucleation seeds.

In addition, to solve the problem of capacity decay with long-term cycling in SnO_2 based anode, the Pd-coated RGO/SnO_2 composite was adopted as electrode material to enhance the capacity and cyclic performance [74]. The interfaces interaction between Pd-coated RGO/SnO_2 composite and binder play an important role on the stability of anode, which exhibits an outstanding energy capacity up to 718 mAh g^{-1} at current density of 100 mA g^{-1} after 200 cycles and good rate performance of 811, 700, 641, and 512 mAh g^{-1} at current density of 100, 250, 500, and 1,000 mA g^{-1}, respectively.

Very recently, ternary metal oxides/GO (RGO) composites were used for anode in LIBs to improve the stability during the discharge-charge cycle in the long term [75, 76]. The electrode of the Zn_2GeO_4–GO nanocomposite synthesized by Zou et al. shows extraordinary high specific capacity, superior rate capability and long cycle life owing to the synergistic coupling of Zn_2GeO_4 and GO layers [76]. Even after 100 discharge-charge cycles, it delivers a specific capacity as high as 1,150 mAh g^{-1} at 200 mA g^{-1}.

Apart from the metal oxides, metal sulfides (CoS_2, FeS and MnS) can be also composited with RGO as anode materials [77–79]. Fei et al. proposed a facile direct-precipitation approach to obtain a FeS-based anode [78]. The FeS@RGO nanocomposite with robust sheet-wrapped structure exhibits better electrochemical performance than isolated FeS nanoparticles because of the smaller particle sizes and the synergetic effects between FeS and RGO sheets, i.e., increased conductivity, shortened lithium ion diffusion path, and effective prevention of polysulfide dissolution.

As an alternation to carbon, silicon emerges as a promising anode material for LIBs owing to its high theoretic capacity of ~4,200 mAh g^{-1} and appropriately low working potential [80]. However, serious pulverization of bulk silicon during cycling limits its cycle life [81]. Composition of Si with GO/RGO is an effective strategy to improve the cycling performance [82–84]. Guo et al. reported a mixture of leaf-like GO and Si NPs as the anode of LIBs [82]. Over 100 cycles, the capacity loss of GO/Si composite electrode is almost negligible. However, the irreversibility on the initial cycling was observed because of its large surface area and surface groups. Later, a novel hierarchical Si nanowire (Si-NW)/RGO composite was reported [84], where the uniform-sized (111)-oriented Si NWs are well dispersed on the Au nanoparticle decorated RGO surface and in between RGO sheets. The Si-NW/RGO composite anode exhibits a highly reversible cycling retention over 100 cycles with a high Li storage capacity of 2,300 mAh g^{-1} at a C/3 rate.

GO/RGO based materials also possess significant advantages compared to the conventional polymer cathodes and lithium-TMOs cathodes in LIBs [85–94]. A theoretical investigation performed by Stournara et al. predicted high-potential lithiation of epoxide on GO/RGO, which suggests a new usage of GO as the cathode material in lithium storage [85]. Later, the epoxide-enriched GO has been demonstrated of capable to be lithiated/delithiated as a rechargeable cathode with high capacity and good stability [86]. Facilitated by the numerous epoxide groups, GO delivers a high capacity of 360.4 mAh g^{-1} at 50 mA g^{-1}. The epoxide-dependent lithiated/delithiated character has been further proved by Ha et al. [87]. Moreover, the hydroxyl groups are identified as the lithiation-active species.

In addition, there have been some efforts on RGO composited with the TMO cathodes in LIBs [89–94]. Enwrapping the $Li_3V_2(PO_4)_3$ into RGO sheets as cathode material can facilitate the charge transfer [89]. However, such $Li_3V_2(PO_4)_3$/RGO composite exhibits a low initial discharge capacities of 170 mAh g^{-1} at 0.1 C rate between 3.0 and 4.8 V. Even with the modification of carbon [89] and cetyltrimethyl ammonium bromide (CTAB) [90] in the $Li_3V_2(PO_4)_3$/RGO composite, the storage capacity only slightly increases to about 170 mAh g^{-1}. For comparison, vanadium oxides/RGO composite exhibits better lithium-storage performance than $Li_3V_2(PO_4)_3$/RGO [91–94]. For RGO/V_2O_5 NWs at a rate of 0.2 C between 2.0 and 4.0 V, the initial and 60th discharge capacities are 225 and 125 mAh g^{-1}, respectively [93].

5.2.2 Lithium-Sulfur Batteries

Lithium-sulfur batteries has a high specific capacity of about 1,675 mAh g^{-1} and a theoretical specific energy of 2,600 Wh kg^{-1} (considering the complete reaction of Li with S to form Li_2S), far beyond the currently matured LIBs. However, the commercial application of Li–S batteries is impeded by several operational problems, such as the insulating nature of sulfur [95], the volume variation and aggregation of insulator Li_2S_2 and Li_2S during the charge-discharge process [96–98], and the capacity fading induced by the high solubility of the polysulfide (shuttle effect) [99]. To alleviate these problems, GO or RGO is proposed to composite with the sulfur cathode. The oxygen-containing functional groups on the basal planes and edges and the structural defects provide the strong anchoring points [100, 101], while the two-dimensional sheet-like structure promotes the formation of self-assembled films for improving the electrochemical performance of Li–S batteries.

Ji et al. [102] applied a low-cost and environmentally benign approach to immobilize sulfur and lithium polysulfides via the reactive functional groups on GO. The GO/S nanocomposite cathodes offer buffering space for the volume change of sulfur during the charging-discharging process. Their ab initio calculations also indicated the role of both epoxy and hydroxyl groups on enhancing the sulfur binding. Moreover, strong interaction between GO and sulfur or polysulfides effectively prevents the active material loss due to the Li polysulfides dissolved in the electrolyte during cycling [102, 103] and therefore leads to a high reversible capacity of 950–1,400 mAh g^{-1}. However, the stable charge-discharge process of pure GO coated sulfur composite can only maintained up to 50 cycles, followed by an obvious capacity fading. The composition of a PEG surfactant-coated GO/S (shown in Fig. 5.12) [104] achieves a stable cycle more than 100

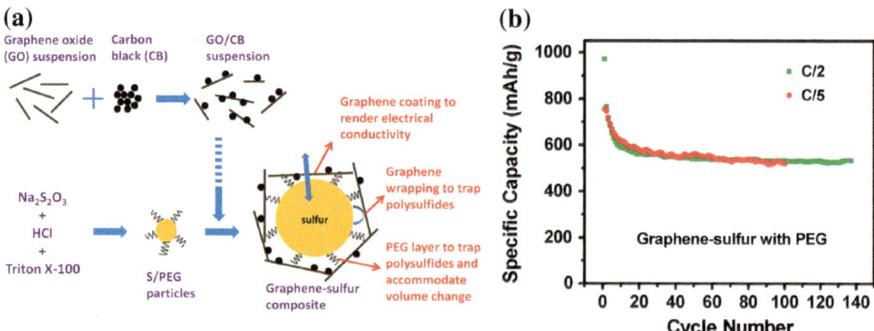

Fig. 5.12 a Schematic of the synthesis steps for a GO/S composite, with a proposed schematic structure of the composite. **b** Cycling performance of the GO/S composite with PEG coating at rates of ~C/5 and ~C/2. Reprinted with permission from Ref. [104]. Copyright (2011) American Chemical Society

times with the compromised capacity drop from initial 1,000 to ~600 mAh g^{-1}. Later attempts of compositing GO with sulfur further convince the contribution of oxygen functional groups on GO to suppression of the shutter effect and the induced good capacity retention [105, 106].

To improve the electrochemical performance of GO/S compositions, the cetyltrimethyl ammonium bromide (CTAB) has been utilized to modify the GO-S cathode [107]. The resulted nanocomposite possesses a high specific capacity of ~800 mAh g^{-1} at 6 C and achieves a long cycle life >1,500 cycles and an extremely low decay rate (0.039 % per cycle). To suppress the Li polysulfides escaped through the open channels among the GO layers, Zhou et al. [108] proposed another strategy of modifying the GO/S nanocomposite with an amylopectin to construct a 3-D cross-linked structure with largely improved cyclability. Additionally, Cui et al. [109] adopted the Li$_2$S as the cathode material advanced by the safety concerns and structure stability during volumetric contraction [110]. The Li$_2$S/graphene oxide composite yields a high discharge capacity of 782 mAh g^{-1} of Li$_2$S (1,122 mAh g^{-1} of S) with stable cycling performance over 150th charge-discharge cycles.

RGO-based sulfur cathode has also received intense attentions since it possesses comparable advantages with GO while prevailed by better electronic conductivity [111–114]. The S@RGO composites with a unique saccule-like structure can provide buffer space to accommodate stress and volumetric expansion of sulfur during the charge-discharge process, which displays a discharge capacity of 724.5 mAh g^{-1} (sulfur mass only) at a current rate of 1 C (1 C = 1675 mA g^{-1}) between 1.2 and 3.0 V [111]. In addition, the Li$_2$S/RGO nanocomposite with a unique 3-D pocket structure and a high initial capacity of 982 mAh g^{-1} was proposed by Han et al. [112]. But the remaining problem of polysulfide dissolution (even with the presence of functional groups on RGO) leads to noticeable capacity fade in the long term [111, 112].

Compared to the multistep synthesis processes with relatively high cost and low synthesis efficiency, the one-step synthesized GO/S cathode exhibits a reversible capacity of 808 mAh g^{-1} at a rate of 210 mA g^{-1} and an average columbic efficiency of ~98.3 % over 100 cycles [113]. The porous GO architecture offers flexible confinement effect that helps prevent the loss of active materials, thus extending the cycling life of the electrodes. Moreover, RGO provides a conductive network surrounding the sulfur particles, which facilitates both electron transport and ion transportation.

Further modification of the GO/S cathode materials can be categorized into different kinds of carbon decoration, such as graphene [115], pristine carbon [116], porous CMK-3 [117, 118], and CNT [119, 120]. A thermally exfoliated graphene-sulfur nanocomposite coupled with RGO for cathode possesses superior electrical conductivity and effective restraint ability to the polysulfides during recycles. The composite delivers a reversible capacity of ~667 mAh g^{-1} after 200 cycles at a high rate of 1.6 A g^{-1}, along with a high Coulombic efficiency of 96 % [115]. The synergistic combination of RGO with mesoporous CMK-3/sulfur composite provides sufficient buffering space and efficient diffusion channel, concomitant with the restrain of polysulfides within cathode, which remarkably alleviates the capacity fade and improve electrochemical properties [117, 118]. Similar synergistic effects of RGO sheets and MWCNTs in sulfur cathode [119, 120] result in excellent electrochemical performance such as larger sulfur loading up to 70 % and higher initial capacity of 1,396 mAh g^{-1} at a current density of 0.2 C (see Fig. 5.13) [119], compared to the case of CMK-3 coated RGO/S composite [117, 118]. Moreover, the ferric chloride has been introduced as not only an oxidizing agent, but also a soft template to form a uniform deposition of sulfur composite, in which sulfur are embedded into a 3-D conducting composite of RGO and polyethylene glycol (PEG) [121]. The synergistic effects of RGO and PEG account for the alleviated high solubility of the polysulfide, and subsequently the high capacity, improved Coulombic efficiency and extraordinarily stable cycling performance.

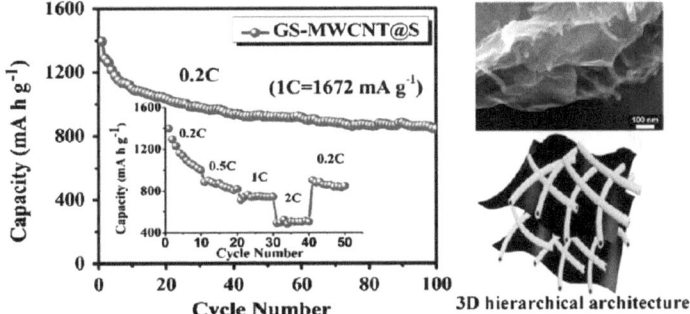

Fig. 5.13 Configuration (*right*) and storage capacity (*left*) of GS-MWCNT@S based electrode in LiS battery. Reprinted with permission from Ref. [119]. Copyright (2013) American Chemical Society

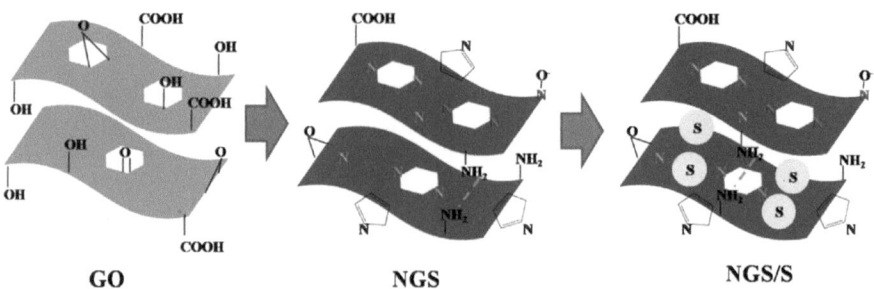

Fig. 5.14 Schematic representations of the synthesis process of NGS/S composites. Reprinted with permission from Ref. [123]. Copyright (2014) Elsevier B.V.

In addition, nitrogen doping in graphene or graphene oxide as electrode is an effective strategy to improve the electrochemical performance of lithium batteries [122]. The N atoms not only improve the electric conductivity of graphene framework but also assist the graphene to immobilize sulfur and confine the diffusion of soluble polysulfides. Wang et al. [123] reported two types of N-doped RGO/S composites, as displayed in Fig. 5.14. Compared with the RGO/S composite, the pyridinic-N enriched RGO/S shows better electrochemical performance, e.g., a remarkably high reversible capacity (1,356.8 mAh g^{-1} at 0.1 C) and long cycle stability (578.5 mAh g^{-1} remaining at 1 C up to 500 cycles). Such improvement can be attributed to the chemical interaction between nitrogen and polysulfide as well as the enhanced electric conductivity of the carbon matrix.

5.2.3 Lithium-Air Batteries

Compared to the Li-ion and Li-S batteries, the Li-O$_2$ batteries depict super promising prospects owing to its extremely high theoretic energy density (up to 13,000 Wh kg^{-1}, comparable to the gas oil) [124, 125] and attract increasing attentions in recent years. There are still more major challenges for the cathode, e.g., the diffusion path blocked by the discharged products of Li$_2$O$_2$ and Li$_2$O [126], the atmospheric moisture induced corrosion [127], the attack by highly reactive reduced O$_2^-$ [128]. Therefore, the ideal cathode materials require good electric conductivity, outstanding O$_2$ reduction performance, structural stability, and suitable path for fast oxygen diffusion. Carbon materials have been used in the Li-O$_2$ cathode due to its superior capability to transport the electrons, support the catalysts and accommodate discharged products for the oxygen reduction reaction (ORR) [129, 130].

The hierarchically GO-based materials with the highly interconnected 3-D channels for rapid oxygen diffusion have been proposed to improve the electrochemical performance of Li-O$_2$ batteries [131, 132]. The electrode with assisted binder materials shows an exceptionally high capacity of 15,000 mAh g^{-1} in Li-O$_2$ batteries [131], whereas the free-standing electrode delivers a high capacity of 11,060 mAh g^{-1} at a current density of 0.2 mA cm^{-2} (280 mA g^{-1}) [132]. Moreover, the defects and functional groups are attributed to the formation of isolated nanosized Li$_2$O$_2$ particles and alleviation of air blocking in the cathode, therefore leading to the improved electrochemical properties [131]. However, a more recent study argued that the functional groups of epoxy and carbonates generated during the discharge reaction will limit the rechargeability of Li-O$_2$ cells [128].

In addition, a novel combination of leaf-like GO and CNT midrib used in cathode of Li-O$_2$ batteries delivers a discharge capacity of 6,000 mAh g$_{GO}^{-1}$ with a cut-off voltage of 2.0 V and a stable cyclic over 150th [82]. In comparison, the discharge capacities are 2,250 mA g$_{GO}^{-1}$ for the GO-based Li-O$_2$ batteries, 2,500 mA g$_{CNT}^{-1}$ for CNT-based Li-O$_2$ battery, and 3,000 mA g$_{CNT/GO}^{-1}$ for CNT/GO mixture-based Li-O$_2$ battery, respectively. Certainly, the synergistic effects of leaf-like GO and CNT contribute to the superior electrochemical properties. Further enhancement of electrochemical performance can be achieved via the nitrogen doped GO as cathode catalyst [133]. The graphitic N content accounts for the limiting current density and the pyridinic N content improves the onset potential for ORR, while the total N content does not play an important role in the ORR process.

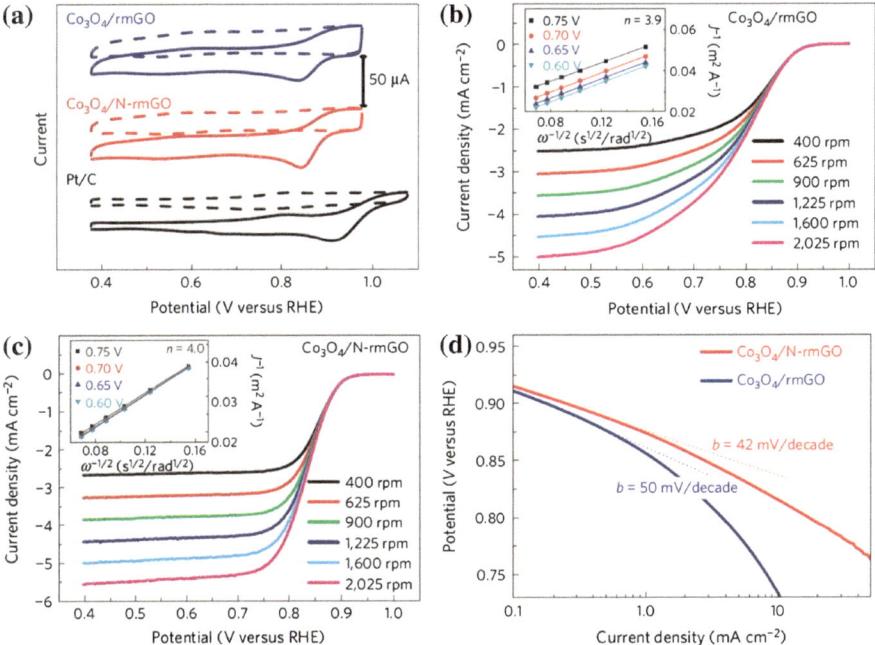

Fig. 5.15 Co$_3$O$_4$/RGO hybrid as oxygen reduction catalysts. Reprinted with permission from Ref. [135]. Copyright (2011) Nature Publishing Group

To improve the electrochemical performance and especially to reduce the high overpotential, efficient catalysts are essential. As an ideal supporter, GO/RGO based composites can achieve a high dispersion and low aggregation of metal catalysts [134–136]. The Ru based nanomaterials supported by RGO show excellent electrocatalytic activity to reduce the high overpotential of oxygen evolution reaction (OER) in Li-O_2 cells [134]. Compared to the Ru based nanocomposites, the MO based nanomaterials hybrid with RGO exhibit superior electrocatalytic activity for the OER reaction. The system proceeds with an average charging potential of ~3.7 V even at a high capacity of 5,000 mAh g^{-1}. However, the stable cycling of the Ru-based materials can only sustain up to 30 cycles due to the degradation of the Li-metal anode, pore clogging, and other issues. In addition, Co and Mn based metal oxides supported on RGO have also been adopted as catalysts in Li-O_2 batteries [135, 136]. As illustrated in Fig. 5.15, the hybrid material by Co_3O_4 nanocrystals and RGO shows high catalytic activity to the ORR, which can be further enhanced by nitrogen doping, while the Co_3O_4 or RGO alone has only little catalytic activity [135].

5.3 Supercapacitors

5.3.1 GO and RGO

Pure carbon materials used in the SCs usually show only low specific capacitance. To overcome this difficulty, GO and RGO with oxygen-containing groups, which has been proved to exhibit pseudocapacitance characteristic, are employed as electrode to replace the conventional carbon materials (e.g., activated carbon, mesoporous carbon, CNT, nanofiber) [137]. In principle, GO can not be used in SCs due to its intrinsically poor electric conductivity. However, Zabihinpour et al. [138] synthesized the multilayer graphene oxide (MGO) nanosheets for supercapacitors, which exhibit 60 F g^{-1} of total electrode material and capacitance retention of 85 % after 10,000 cycles. In addition, Shulga et al. [139] indicated that the deeply oxidized graphene oxide can be used as a separator in supercapacitors composed of polyaniline thin film electrodes, since it has a proton type conductivity after being permeated with a solution of sulfuric acid. This supercapacitors achieves to about 150 F g^{-1} based on the total weight of whole system (electrode, separator, and electrolyte).

To improve the electric conductivity of GO and retain the pseudo-capacitive behavior, partial reduced GO has been adopted [140–152]. A superior capacitance of functionalized graphene (fG) prepared by a solvothermal method was reported to exhibit a specific capacitance up to 276 F g^{-1} at 0.1 A g^{-1} in a 1 M H_2SO_4 electrolyte with good rate performance and cycling stability, which was ascribed to the surface oxygen containing groups (mainly carbonyls and hydroxyls) induced large pseudocapacitance, less aggregation, and good wetting properties [140]. In addition, Zhang et al. [141] used a water-soluble RGO as electrode materials in SCs with a high specific capacitance of 238 F g^{-1} at the current density of 0.1 A g^{-1} in 1 M H$_2$SO$_4$ electrolyte, which is the co-contribution of double layer capacitance and pseudocapacitance from oxygen-containing groups.

The pH value has an important effect on the performance of supercapacitors [142]. RGO sheets in the acidic and neutral media have more oxygen-functional groups and lower surface areas but broader pore size distributions (both EDL and pseudocapacitive behavior) than those in the basic medium (mainly EDL), thus leading to higher specific capacitance of 225 F g^{-1} (acidic) and 230 F g^{-1} (neutral), respectively, compared to 185 F g^{-1} for the case of basic medium at a constant current density of 1 A g^{-1}.

The reduction level of the graphene sheets also plays a significant role in controlling the intrinsic properties such as the interlayer spacing, oxygen content, BET specific surface area [143, 144]. Among these effects, the variation of oxygen-containing groups is the main factor influencing the overall performance of SCs. Increasing the reduction time may facilitate removal of the oxygen-containing group, which results in two compromising effects on the specific capacitance of the RGO-based SCs: (a) loss of pseudocapacitance; (b) increase of the electric conductivity and thus enhanced electric double-layer capacitance as well as recovery of aromatic carbon rings and decrease of the defect density.

Moreover, chemical doping of RGO can tune the electronic properties to enhance the supercapacitor performance. B-doped reduced graphene oxide (B-RGO) with a high specific surface area of 466 m^2/g, exhibits excellent supercapacitor performance including a high specific capacitance of 200 F g^{-1} in aqueous electrolyte and good stability over 4,500 cycles [145]. On the other hand, N-doped RGOs also act as electrode in SCs [146–148] and the configuration of N atoms on RGO has remarkable influence on the overall supercapacitor performance [148].

Despite the progresses in GO/RGO based SCs, RGO usually encounter the self-aggregation tendency at the presence of chemically bonded water molecules intercalated between the graphene layers [153–155], which dramatically decrease the specific surface area of electrodes, impedes the ion diffusion from the electrolyte to the electrodes, and reduces the effective capacitance [156]. It is thus necessary to develop the non-stacking GO/RGO architecture to increase the surface area and the pore volume between the non-stacked graphene layers simultaneously. Considering the tremendous influence of the intercalated water molecules on restacking of the RGO at the early dried GO stage, Lee et al. [149] functionalized GO sheets with melamine resin (MR) monomers to prevent the hydrogen bonding with water molecules upon drying and consequently obtained the porous restacking-inhibited GO sheets with specific capacitance of 210 F g^{-1}, superiors capacitance retention for 20,000 cycles and ~7 s rate capability. Moreover, the anti-solvent (hexane) without any interaction with the various oxygen groups of GO has been used to replace the water solution for fabricating the non-stacked RGO via a simple, economical and effective way [150], as shown in Fig. 5.16. The obtained non-stacked RGO possesses a high

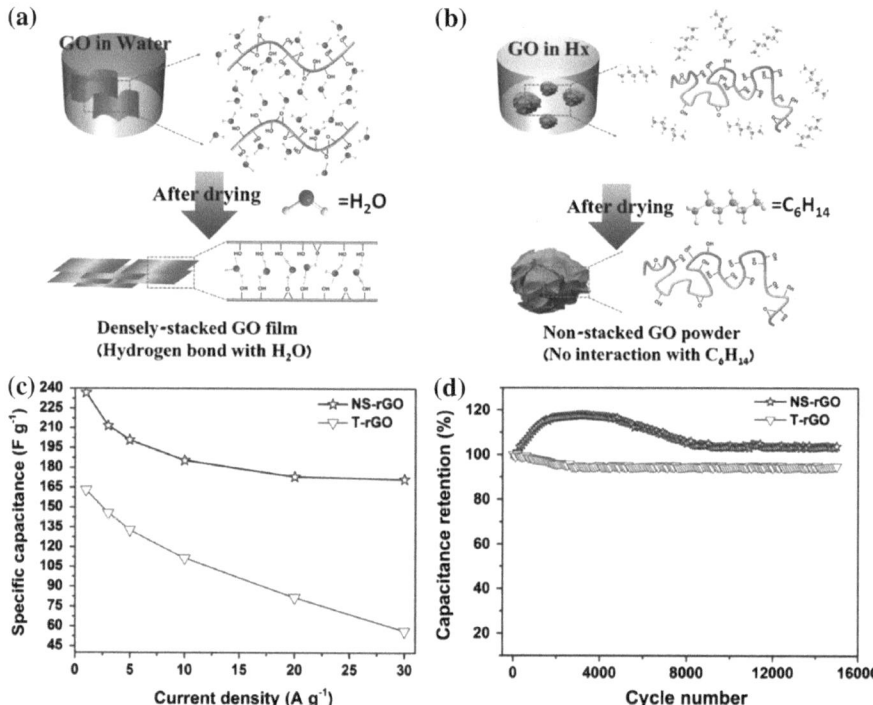

Fig. 5.16 Schematic illustration of **a** a densely-stacked GO film formation in water; **b** GO in hexane, named as non-stacked GO (NSGO). **c** Dependence of specific capacitance on current density in thermally reduced GO (T-rGO) and NS-rGO based supercapacitors with 6.0 M KOH electrolyte. **d** Cyclic stability and retention of T-rGO and NS-rGO. Reprinted with permission from Ref. [150]. Copyright (2013) WILEY-VCH Verlag GmbH & Co. KGaA, Weinheim

surface area with unique structural properties and consequently presents a high specific capacitance of up to 236.8 F g^{-1} at a current density of 1 A g^{-1} and a long cycle life.

On the other hand, incorporation of Mg(OH)$_2$ nanosheets as a "spacer" and template in-between GO sheets is an effective strategy not only to inhibit the restacking behavior during the heat-treatment of GO, but also to preserve the stable oxygen-containing groups [151]. As illustrated in Fig. 5.17, the relationship between thermal temperature and pseudocapacitance indicates the significant role of oxygen containing groups in enhancing the pseudocapacitance, which combines with the high surface area and powder density, co-contributing to the ultrahigh specific gravimetric and volumetric capacitances of 456 F g^{-1} and 470 F cm^{-3}, almost 3.7 times and 3.3 times higher than hydrazine reduced GO, respectively. Recently, Maiti et al. [152] presented a simple way to fabricate the graphene gel consisting of highly conductive 3-D RGO framework for SCs. The open pores within the RGO gel facilitate the aqueous electrolyte transport and achieve a large area capacity without sacrificing rate capability.

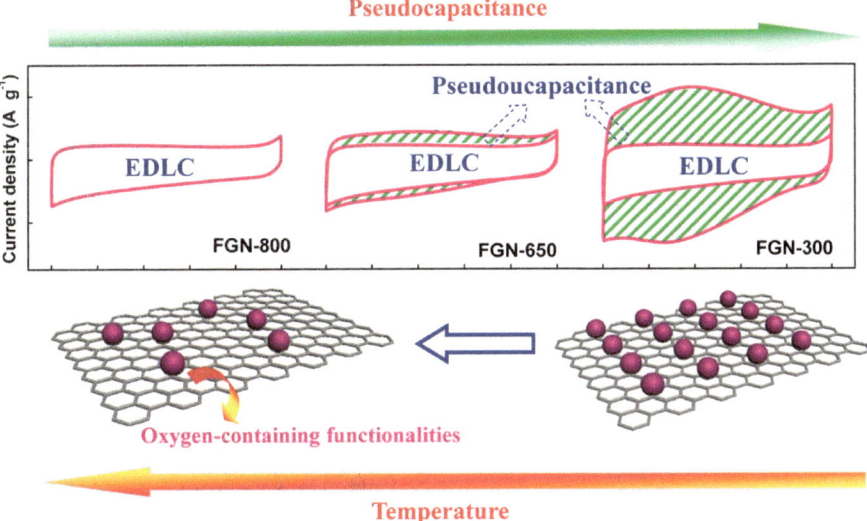

Fig. 5.17 Illustration of the relationship between thermal temperature and pseudocapacitance. Reprinted with permission from Ref. [151]. Copyright (2014) American Chemical Society

5.3.2 GO/RGO Combined with Other Carbon Materials

Modifying the conventional carbon material with GO/RGO is a feasible strategy to enhance the supercapacitance [157–164]. To solve the issue of intrinsic insolubility of CNTs in water, GO with high solubility has been introduced as surfactant to disperse CNTs through strong π–π stacking interaction [165–167]. Dong et al. [157] have experimentally demonstrated the spontaneous formation of SWCNT/ GO core-shell structure. But its application in the supercapacitors has been limited since the highly conductive SWCNTs are completely wrapped by the insulating GO nanosheets. Therefore, the SWCNT/GO sample has been further reduced to SWCNT/RGO, which shows a specific capacitance of 194 F g^{-1}, compared to the SC electrodes with RGO (155 F g^{-1}) and SWCNT (127 F g^{-1}). Using MWCNT as core and GO nanoribbon (GONR) as shell, a core-shell MWCNT@GONR structure for supercapacitor application was reported to possess a high specific capacitance of 252.4 F g^{-1}, higher than 194 F g^{-1} for the SWCNT/GO electrode [158].

In addition, intercalation of CNT as the spacer into the GO layers is an effective way to inhibit the GO restacking [168–170]. Zeng et al. [159] proposed to incorporate the super-short CNT (SSCNT) into RGO layers to synthesize the 3-D multilayer RGO/SSCNT architecture for supercapacitors, as shown in Fig. 5.18. This strategy not only increases the utilization of the closed pore volumes of MWCNTs but also inhibits the aggregation of RGO. Simultaneously, the open-end pipes provide abundant transport paths and short axial transmission distance for the electrolyte ions and electrons in the electrode. As a consequence, the RGO/SSCNTs

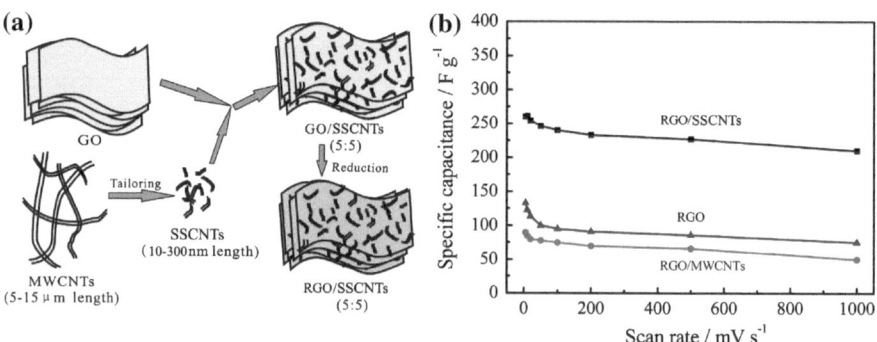

Fig. 5.18 a Schematic illustration of the formation steps of the multilayer RGO/SSCNT architecture. **b** Specific capacitance against CV scan rate of RGO/SSCNTs, RGO and RGO/MWCNTs. Reprinted with permission from Ref. [159]. Copyright (2013) Elsevier B.V.

display a high specific capacitance of 210 F g^{-1} even at 1,000 mV s^{-1}, compared to the values of 75 F g^{-1} for RGO and 50 F g^{-1} for RGO/MWCNTs. Similarly, Beidaghi et al. [160] used the CNTs as the spacer and fabricated a micro-patterned interdigitated electrodes based on RGO-CNT composites for SC application. The interdigitated RGO-CNT electrodes can increase the accessibility of electrolyte ions into graphene sheets and therefore lead to a high-frequency response of the micro-supercapacitors with resistive-capacitive time constants as low as 4.8 ms. Furthermore, the GO/RGO-CNT composites have been deposited on the carbon cloth for SC application [161, 162]. In these composites, GO plays an important role in facilitating the deposition of CNTs on carbon cloth, and the existence of CNTs prevents the GO restacking simultaneously. These synergistic effects co-contribute to the significantly enhanced supercapacitor performance [161].

Apart from the CNT, other carbon materials have also been combined with GO/RGO for SC application [163, 164]. The electrochemically exfoliated graphene (G) and wet-chemically produced GO have been combined as G/GO/G, where the GO acts as the dielectric spacer to enhance the supercapacitance [163]. Meanwhile, the electrochemically reduced graphene oxide (ERGO) coated carbon fibers (CFs) have been used as supercapacitor electrodes, which exhibit superior electrical charge storage performance [164]. The addition of ERGO results a unique wrinkled yet porous structure with higher available surface area for accumulating the electrolyte ions to produce excellent electrochemical double-layers.

5.3.3 GO/RGO-Transition Metal Oxide Composites

The pseudo-capacitors with the TMOs and conducting polymers as electrode are restricted by the low working voltage, poor stability and unsatisfactory high-rate capabilities [171, 172]. In principle, the concept of synergistic composition of nanostructured conducting polymers and metal oxides with GO/RGO electrode material has been demonstrated to be feasible to possess the pseudo-capacitance and EDLC behavior and to improve the overall supercapacitor performance simultaneously [173]. However, in practical situations, overall supercapacitor performance varies with the specific composition. For example, the GO/manganese oxide based SCs show great electrochemical performance [174–176]. The composite of GO supported by needle-like MnO_2 nanocrystals (see Fig. 5.19) exhibits a specific capacitance of 197.2 F g^{-1}, slightly lower than the value of 211.2 F g^{-1} for pure nano–MnO_2 [174]. However, this small depression has been compensated by the enhanced electrochemical stability. After 1,000 cycles, the CMG_{15} (chemically synthesized GO/MnO_2 nanocomposites when the feeding ratio of MnO_2/GO is 15:1) electrode retains about 84.1 % (165.9 F g^{-1}) of initial capacitance, while the nano-MnO_2 one retains only about 69.0 % (145.7 F g^{-1}). Later, the GO/Mn_3O_4 composite was reported to possess a high specific capacitance of 344 F g^{-1}, a specific energy of 93 Wh kg^{-1}, and simultaneously a long cycling life time beyond 3,000th [175]. The enhanced supercapacitor performance of GO/Mn_3O_4 composite is attributed to the GO induced surface structure modification, strong trapping capability of Mn_3O_4 particles during the fast charging-discharging, and the accessibility of electrolyte ions.

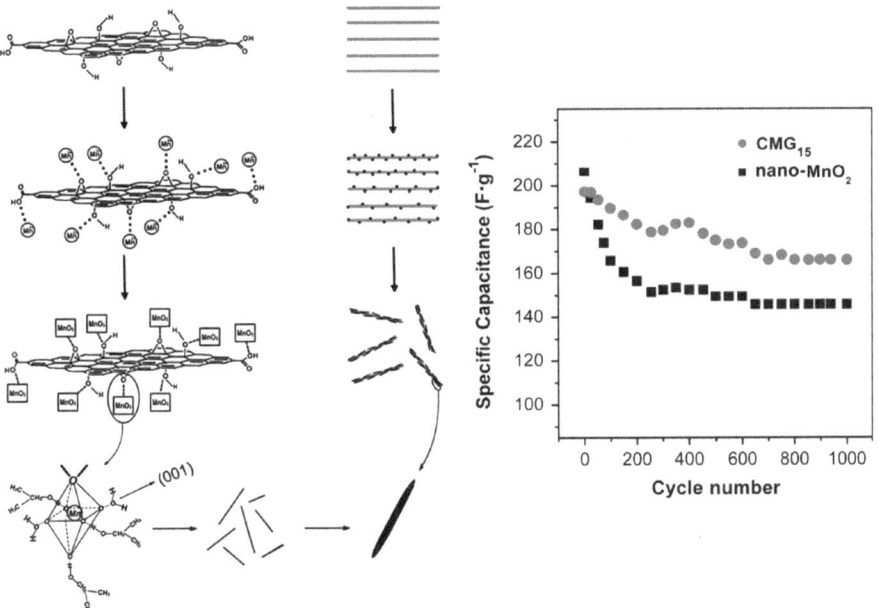

Fig. 5.19 Formation mechanism for GO/MnO_2 nanocomposites. Reprinted with permission from Ref. [174]. Copyright (2010) American Chemical Society

Despite the above advances, the major limitation of poor electric conductivity of GO remain exists. In this regard, manganese oxide/RGO composites with controllable functional groups may possess better electric conductivity and thus better supercapacitor performance [177–179]. The reduction level plays a significant role in the final electrochemical and supercapacitor performance. Li et al. [177] reported a graphene/MnO_2 composite for supercapacitor electrodes prepared via hydrazine reduction of the graphite oxide and a soft chemical precipitation route at relatively low temperature. The MnO_2/RGO composite exhibits a maximum specific capacitance of 124 F g^{-1} at 200 mA g^{-1} with 1 M Na_2SO_4 as the electrolyte. The resulted performance is ascribed to the compromising contributions by the oxygen-containing groups: (1) they may increase the resistance of the composites; (2) they benefit the dispersion of MnO_2 on graphene and the infiltration of electrolyte. Furthermore, the MnO_2/H-RGO and MnO_2/S-RGO electrodes have been fabricated by the chemical reduction with the reducing agents of hydrazine hydrate and sodium borohydride, respectively [178]. These electrodes display high specific capacitance (327.5 F g^{-1} for H-RGO/MnO_2 and 278.6 F g^{-1} for S-RGO/MnO_2 at a scan rate of 10 mV s^{-1}) and good cyclic stability up to 1,000 cycles.

Moreover, RGO composited with other TMOs have been reported [180–191]. A unique hybrid architecture of Co_3O_4/RGO scrolls is prepared through a two-step surfactant assisted method, which enables almost every single Co_3O_4 scroll to connect with the RGO platelets through the residual oxygen-containing groups and thus integrate both active material and "conducting matrix" [180]. The Co_3O_4/RGO scrolls show a remarkable electrochemical performance in terms of a specific capacitance of 163.8 F g^{-1} at a current density of 1 A g^{-1} and better capacitance retention of 93 % than that of Co_3O_4 (91 %) after 1,000 cycles.

Later, a simple one-step synthetic strategy has been developed to prepare the Co_3O_4 NSs–RGO hybrid for SC application [182]. The Co_3O_4 NSs–RGO hybrid with 7.2 wt% Co_3O_4 loading yields higher specific capacitance of 187 F g^{-1} at 1.2 A g^{-1} than the case of Co_3O_4/RGO scrolls. Furthermore, the capacitance depresses by only 6–9 wt% at constant current densities of 1.2 and 5 A g^{-1} after 1,000 continuous charge-discharge cycles. The synergetic effects of nanoscale size and good redox activity of the Co_3O_4 NPs combined with the high electric conductivity of the RGO result in the enhanced electrochemical utilization at high rates.

As for the iron oxides, the specific capacitance of the Fe_3O_4/RGO nanocomposites is highly dependent on the mass ratio of Fe_3O_4 and RGO [183, 184]. The nanocomposite synthesized by a one-step solvothermal process with the Fe_3O_4:RGO ratio of 2.8 shows highest specific capacitances of 480 F g^{-1} at the discharge current density of 5 A g^{-1} with the corresponding energy density of 67 Wh kg^{-1} at a power density of 5,506 W kg^{-1} [183]; and an extremely high specific capacitance of 843 F g^{-1} is obtained at a discharge current density of 1 A g^{-1} with the corresponding energy density of 124 Wh kg^{-1} at a power density of 332 W kg^{-1}. Moreover, the Fe_3O_4/RGO nanocomposites show stable cycling

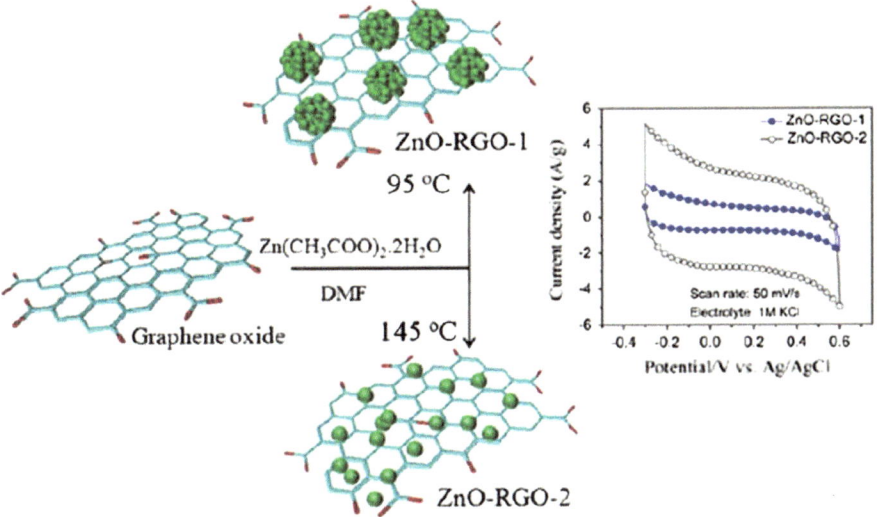

Fig. 5.20 Reaction temperatures on the morphologies of zinc oxide-reduced graphene oxide (ZnO–RGO) nanohybrids and the cyclic performance. Reprinted with permission from Ref. [189]. Copyright (2014) American Chemical Society

performance without any lose in the specific capacitance after 1,000 charge-discharge cycles.

The nickel oxides/RGO composites also possess superior electrochemical performance as SC electrodes [186–188]. A 3-D flower-like hierarchical NiO/RGO composite has been fabricated for supercapacitor applications, in which the NiO NPs not only inhibit the RGO restacking but also function as catalyst for the GO reduction [187]. The hybrid composite electrode endows great supercapacitor performance, that is, a maximum specific capacitance up to 428 F g^{-1} at a discharge current density of 0.38 A g^{-1} in a 6.0 M KOH electrolyte is achieved. Moreover, Prakash et al. [189] pointed out that, among the several GO induced effects, the morphology control for ZnO nanoparticles is the key factor to determine the overall performance of the electrodes (see Fig. 5.20). Consistent conclusion has been drawn in NiO/RGO nanohybrids [188].

Apart from the above mentioned metal oxides, CuO/RGO hybrid lamellas exhibit a specific capacitance of 163.7 F g^{-1} (much higher than 69.7 F g^{-1} of CuO nanosheets and 66.0 F g^{-1} of RGO, respectively), which is dominated by the synergistic effects of redox activity of the CuO nanosheets, the good electric conductivity of the RGO, and the unique sandwich-like porous structures spaced by CuO nanospheres [190]. TiO$_2$ nanobelts composited with RGO (RGO/TiO$_2$ NBs) used as electrode in SCs show better supercapacitor performance than the case of TiO$_2$ nanoparticles [191]. At a discharge current density of 0.125 A g^{-1}, the RGO/TiO$_2$ NBs yields a much higher specific capacitance of 225 F g^{-1} than 62.8 F g^{-1} for RGO-TiO$_2$ NPs.

5.3.4 Composites of GO/RGO and Conducting Polymers

Compared to the TMOs, composites by conducting polymers and GO/RGO show even higher supercapacitance and have also been extensively documented [192–212]. The composition mechanisms of GO/RGO and the conducting polymers influence the supercapacitor performance.

Among all polymers, polyaniline (PANI) is a promising material owing to its high capacitive characteristics, low cost, and ease to synthesize [213, 214]. However, the relatively poor cycling life limits its practical applications. GO with high surface area and oxygen-containing groups has excellent capability to highly disperse the PANi and thus form stable hybrid composite for SC electrodes [192–195]. The hierarchical nanocomposites of PANI/GO, in which the PANI nanowire arrays are aligned vertically on GO substrate, show a remarkable synergistic effect of improving the specific capacitance to 555 F g^{-1} (298 F g^{-1} for PANI only) at a discharge current density of 0.2 A g^{-1} and retaining the capacitance to 92 % (74 % for PANI only) after 2,000 charge-discharge cycles [193]. Wang et al. [194] synthesized a PANI/GO composite and proposed their possible interaction mechanisms, including electrostatic interaction, hydrogen bonding and π-π stacking (see Fig. 5.21). The composite used in SC shows maximum initial specific capacitances of 746 F g^{-1}. Unlike the use of the edged carboxyl groups of graphene sheets as linkers, to fully use the ample oxygen-containing groups on the basal plane of graphene sheets, Liu et al. [195] performed oxalic acid treatment to achieve the carboxyl-functionalized GO (CFGO) for combining the amine-nitrogens. The CFGO/PANI composite delivers a specific capacitance of 525 F g^{-1}, twice of the reduced CFG–PANI (262 F g^{-1}) at a current density of 0.3 A g^{-1}. Furthermore, CFGO/PANI electrode keeps higher capacitance retention of 91 % than 85 % for CFG/PANI. The lower capacitive performance of CFG–PANI composite is resulted from the absence of structure destruction induced by oxygen-containing functional groups, which indicates the importance of tuning surface chemistry of GO/conducting polymer composites for supercapacitor performance.

The RGO-PANI composites have been further developed [196–199]. A polyaniline-grafted reduced GO (PANI-g-RGO) composites were prepared in a three-step process starting from functionalizing GO with 4-aminophenol via acyl chemistry, where a concomitant reduction of GO takes place for supercapacitor application [196]. However, its low electrical conductivity of 8.66 S/cm results in an inadequate capacitance of 250 F g^{-1} and poor cycling stability. To explore the influence of graphene surface chemistry on the supercapacitor performance of graphene/polyaniline composites, four different surface-functionalized graphenes, i.e., GO, RGO, nitrogen-doped RGO (N-RGO), and amine-modified RGO (NH$_2$-RGO), were adapted to composite PANI [197]. Among them, the NH$_2$-RGO/PANI

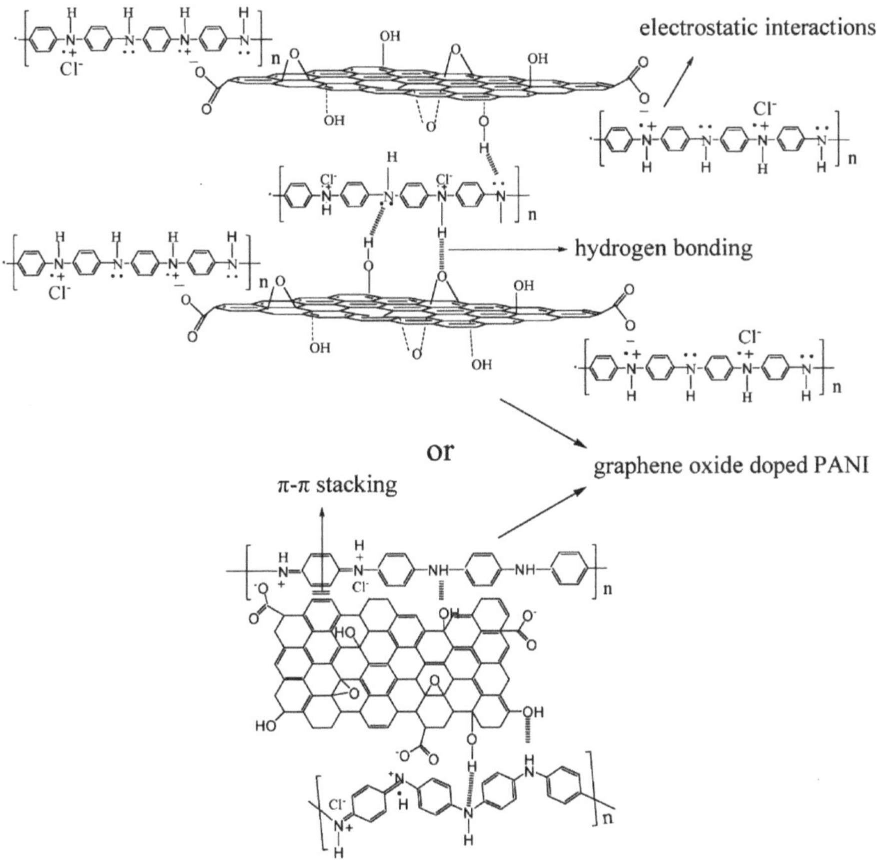

Fig. 5.21 Proposed possible combining modes of GO/PANI composite. Reprinted with permission from Ref. [194]. Copyright (2010) American Chemical Society

composite shows the largest specific capacitance of 500 F g^{-1} and good cyclability with no loss of capacitance over 680 cycles. Recently, Wang et al. [198] reported a hierarchical nanocomposite of PANI-fRGO, in which nitrophenyl groups (initially grafted on RGO via C–C bond) are reduced to aminophenyl to act as anchor sites for the growth of PANI arrays on RGO. The PANI-fRGO nanocomposite based SC show a high specific capacitance of 590 F g^{-1} at a current density of 0.1 A g^{-1} without any loss of capacitance after 200 cycles.

Polypyrrole (PPy) is another kind of attractive conducting polymer in SC application due to its low cost, easy to synthesize and relatively high conductivity [215]. To further improve the performance (especially the cycle stability), GO/RGO-PPy composite have been developed as the electrode materials [200–204]. A composite of PPy nanowires and GO nanosheets is fabricated as the electrode and yields a high specific capacitance of 633 F g^{-1} at a current density of 1 A g^{-1}, which is much larger than that of each individual component [200]. Meanwhile, the capacitance retaining of 94 % for the GO/PPy composite after 1,000 charge-discharge cycles is much more favored than that of 68 % for the pure PPy.

Later, another GO/PPy nanocomposite was fabricated using one-step coelectrodeposition via the synergistic effects that the relatively large anionic GO serves as a weak electrolyte and is entrapped in the PPy nanocomposite during the electrochemical polymerization as well as acts as an effective charge balancing dopant within the PPy [201]. This GO/PPy nanocomposite used as electrode yields a specific capacitance of 356 F g^{-1} at a discharge current of 0.5 A g^{-1} and retains 78 % of the initial capacitance after 1,000 cycles. In addition, one kind of GO/PPy nanowire was prepared via the assistant of Al$_2$O$_3$ template, in which the PPy is electrochemically doped by oxygen-containing groups of the GO [202]. The vertically aligned GO/PPy nanowires with large surface area and intimate contact between the nanowires and the substrate electrode exhibit an extremely high capacitance of 960 F g^{-1} and stable cyclic up to 300th. To further improve the thermal stability and capacitance retention, a combination of RGO and thermally carbonized PPy, was subsequently synthesized but resulted in a decreased capacitance of 200 F g^{-1} without the redox reactions of PPy.

To solve the problem of poor solubility in water and little anionic characteristics for RGO, Park et al. [204] reported a composite of PPy[RGO-PSS] via the PSS (polystyrenesulfonate) as the bridge between PPy and RGO to facilitate charge transfer, as illustrated in Fig. 5.22. This architecture delivers a specific capacitance of 280 F g^{-1} at a high discharge rate of 1,000 C (equivalent to 50 A g^{-1}). The GO/PPy composites have also been demonstrated to be an efficient organic current collector in an all-plastic SC [205].

Composition of GO/RGO with other polymers has also been documented in SC applications [206–212]. The GO-doped ion gel as electrolyte in the all-solid-state supercapacitor exhibits smaller internal resistance, higher capacitance performance, and better cycle stability compared to the case of pure ion gel and

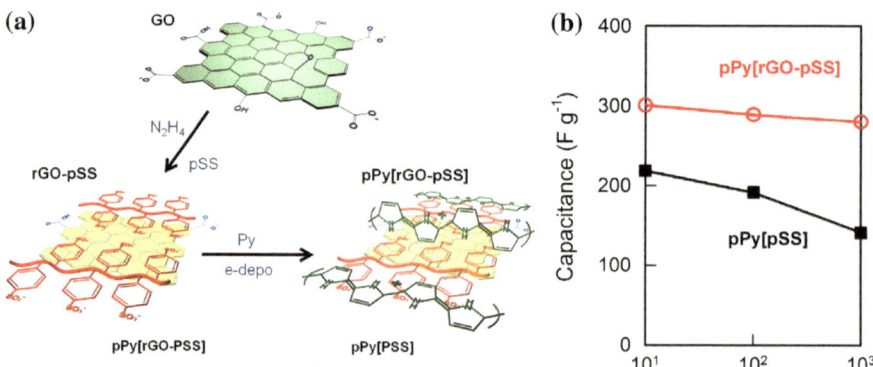

Fig. 5.22 Schematic of PPy[RGO-PSS] obtained by reducing GO in presence of PSS and then electropolymerizing pyrrole in presence of RGO-PSS. Reprinted with permission from Ref. [204]. Copyright (2014) Elsevier Ltd.

the conventional supercapacitor [206, 207]. In addition, the porous GO-based benzimidazole-crosslinked network (GO/BIN) composites display a specific capacitance up to 370 F g^{-1} (partial contributed by the imidazole rings induced pseudocapacitive redox reaction) at a current density of 0.1 A g^{-1} and retain 90 % of the original capacitance after 5,000 cycles at a current density of 3 A g^{-1} [209]. Moreover, RGO/nanofiber used as electrodes [211] and RGO/ionic liquid used as the electrolyte [212] in SCs also exhibit excellent electrochemical performance.

5.3.5 Other GO/RGO-Based Composites

Coupling the promising pseudocapacitive materials of metal sulfides with GO/RGO is another efficient strategy to improve the supercapacitor performance [216, 217]. The synergistic combination of WS$_2$/RGO hybrids exhibit enhanced supercapacitor performance with specific capacitance of 350 F g^{-1} at a scan rate of 2 mV/s (about 5 and 2.5 times higher than bare WS$_2$ and RGO sheets) and energy density of ~49 Wh kg^{-1}, and simultaneously keeping stable over 1,000 cycles [216]. Even more impressively, the NiS/GO nanocomposite display a very high specific capacitance of 800 F g^{-1} at 1 A g^{-1} and long cycle life over 1,000 cycles [217]. On the other hand, the Sb/RGO composite also shows promising potential for supercapacitor electrode due to its high specific capacitance of 289 F g^{-1} (mainly contributed by the pseudocapacitance) and very good cyclability up to 1,000th [218]. The cellulose is also proposed to couple RGO for SC electrode but the composite only delivers a low specific capacitance of 71.2 F g^{-1} at the current density of 50 mA g^{-1} [219].

Based on the GO/RGO-polymer composites, other active materials have been further incorporated into the multiplex architecture for SC electrodes [220–224]. The p-phenylenediamine/GO/Au composite yields an increased specific capacitance of 238 F g^{-1} compared to 11 F g^{-1} of individual PPD and 176 F g^{-1} of PPD/GO and retains 81 % of initial capacitance after 500 cycles [220]. Carbon materials can further hybrid with GO/RGO polymers. The hierarchical freestanding CNF/GO/PANI based SC electrodes possess the 3-D open channels for the diffusion of electrolyte ions and thus exhibit a remarkable specific capacitance of 450.2 F g^{-1} at the scan rate of 10 mV/s and good cyclic stability over 1,000th [221]. Furthermore, an all-solid-state asymmetric supercapacitor using the RGO/MWCNT as the negative electrode and CFP/PPy as the positive electrode only exhibits a specific capacitance of 82.4 F g^{-1} at a current density of 0.5 A g^{-1} but it possesses a stable potential window of 1.6 V and keeps excellent cycling stability up to 2,000 cycles (93 % retention of the initial specific capacitance) [222].

The GO-based ternary composites, simultaneously integrated with the conducting polymers and TMOs, have been used as SC electrodes and exhibit superior electrochemical performance [223, 224]. Based on the GO/RGO-metal oxides, addition of the ternary carbon material may also improve the supercapacitor performance [225, 226]. The MnO$_2$-RGO/CFP composite delivers a specific capacitance of 393 F g^{-1} at 0.1 A g^{-1} (2.2 times higher than the MnO_2/CFP), along with excellent capacity retention of 98.5 % after 2,000 cycles. Moreover, the hydroxides based GO/RGO composites as an extraordinary SC electrode have been addressed [227–229]. A 3-D activated RGO nanocup/Ni-Al layered double hydroxides composite exhibits an extremely high specific capacitance of 2,712.7 F g^{-1} at the current density of 1 A g^{-1} [227]. Even when the current density rises up to 50 A g^{-1}, the specific capacitance remains 1,174 F g^{-1} and keeps 98.9 % of initial capacitance after 5,000 cycles at the current density of 30 A g^{-1}.

References

1. Yeh, T.F., Chen, S.J., Yeh, C.S., Teng, H.: J. Phys. Chem. C **117**, 6516–6524 (2013)
2. Yeh, T.F., Syu, J.M., Cheng, C., Chang, T.H., Teng, H.: Adv. Funct. Mater. **20**, 2255–2262 (2010)
3. Krishnamoorthy, K., Mohan, R., Kim, S.J.: Appl. Phys. Lett. **98**, 244101–244103 (2011)
4. Yeh, T.F., Teng, H.: ECS Trans. **41**, 7–26 (2012)
5. Yeh, T.F., Chan, F.F., Hsieh, C.T., Teng, H.: J. Phys. Chem. C **115**, 22587–22597 (2011)
6. Yeh, T.F., Teng, C.Y., Chen, S.J., Teng, H.: Adv. Mater. **26**, 3297–3303 (2014)
7. Matsumoto, Y., Koinuma, M., Ida, S., Hayami, S., Taniguchi, T., Hatakeyama, K., Tateishi, H., Watanabe, Y., Amano, S.: J. Phys. Chem. C **115**, 19280–19286 (2011)
8. Jiang, X., Nisar, J., Pathak, B., Zhao, J., Ahuja, R.: J. Catal. **299**, 204–209 (2013)
9. Wang, L., Sun, Y., Lee, K., West, D., Chen, Z., Zhao, J., Zhang, S.B.: Phys. Rev. B **82**, 161406–161409 (2010)
10. Wang, L., Lee, K., Sun, Y.Y., Lucking, M., Chen, Z., Zhao, J.J., Zhang, S.B.: ACS Nano **3**, 2995–3000 (2009)
11. Agegnehu, A.K., Pan, C.J., Rick, J., Lee, J.F., Su, W.N., Hwang, B.J.: J. Mater. Chem. **22**, 13849–13854 (2012)
12. An, X., Jimmy, C.Y.: RSC Adv. **1**, 1426–1434 (2011)
13. Osterloh, F.E.: Chem. Mater. **20**, 35–54 (2007)
14. Lv, X.J., Zhou, S.X., Zhang, C., Chang, H.X., Chen, Y., Fu, W.F.: J. Mater. Chem. **22**, 18542–18549 (2012)
15. Xiang, Q., Yu, J.: J. Phys. Chem. Lett. **4**, 753–759 (2013)
16. Iwase, A., Ng, Y.H., Ishiguro, Y., Kudo, A., Amal, R.: J. Am. Chem. Soc. **133**, 11054–11057 (2011)
17. Zhang, N., Zhang, Y., Xu, Y.J.: Nanoscale **4**, 5792–5813 (2012)
18. Li, Q., Guo, B., Yu, J., Ran, J., Zhang, B., Yan, H., Gong, J.R.: J. Am. Chem. Soc. **133**, 10878–10884 (2011)
19. Zhang, J., Yu, J., Jaroniec, M., Gong, J.R.: Nano Lett. **12**, 4584–4589 (2012)
20. Xiang, Q., Yu, J., Jaroniec, M.: Nanoscale **3**, 3670–3678 (2011)
21. Lv, X.J., Fu, W.F., Chang, H.X., Zhang, H., Cheng, J.S., Zhang, G.J., Song, Y., Hu, C.Y., Li, J.H.: J. Mater. Chem. **22**, 1539–1546 (2012)
22. Wang, J., An, C., Liu, J., Xi, G., Jiang, W., Wang, S., Zhang, Q.H.: J. Mater. Chem. A **1**, 2827–2832 (2013)
23. Xiang, Q., Yu, J., Jaroniec, M.: J. Phys. Chem. C **115**, 7355–7363 (2011)
24. Yang, J., Zeng, X., Chen, L., Yuan, W.: Appl. Phys. Lett. **102**, 083101–083104 (2013)
25. Tran, P.D., Batabyal, S.K., Pramana, S.S., Barber, J., Wong, L.H., Loo, S.C.J.: Nanoscale **4**, 3875–3878 (2012)
26. Mukherji, A., Seger, B., Lu, G.Q., Wang, L.: ACS Nano **5**, 3483–3492 (2011)
27. Hou, J., Wang, Z., Kan, W., Jiao, S., Zhu, H., Kumar, R.: J. Mater. Chem. **22**, 7291–7299 (2012)

28. Khan, Z., Chetia, T.R., Vardhaman, A.K., Barpuzary, D., Sastri, C.V., Qureshi, M.: RSC Adv. **2**, 12122–12128 (2012)
29. Xiang, Q., Yu, J., Jaroniec, M.: J. Am. Chem. Soc. **134**, 6575–6578 (2012)
30. Dong, C., Li, X., Jin, P., Zhao, W., Chu, J., Qi, J.: J. Phys. Chem. C **116**, 15833–15838 (2012)
31. Kudo, A., Miseki, Y.: Chem. Soc. Rev. **38**, 253–278 (2009)
32. Patchkovskii, S., Tse, J.S., Yurchenko, S.N., Zhechkov, L., Heine, T., Seifert, G.: Proc. Natl. Acad. Sci. U.S.A. **102**, 10439–10444 (2005)
33. Kim, B.H., Hong, W.G., Yu, H.Y., Han, Y.K., Lee, S.M., Chang, S.J., Moon, H.R., Jun, Y., Kim, H.J.: Phys. Chem. Chem. Phys. **14**, 1480–1484 (2012)
34. Kim, J.M., Hong, W.G., Lee, S.M., Chang, S.J., Jun, Y., Kim, B.H., Kim, H.J.: Int. J. Hydrogen Energy **39**, 3799–3804 (2014)
35. Zhao, Y.F., Kim, Y.H., Dillon, A.C., Heben, M.J., Zhang, S.B.: Phys. Rev. Lett. **94**, 15504–15507 (2005)
36. Sun, Q., Wang, Q., Jena, P., Kawazoe, Y.: J. Am. Chem. Soc. **127**, 14582–14583 (2005)
37. Wang, L., Yang, F.H., Yang, R.T., Miller, M.A.: Ind. Eng. Chem. Res. **48**, 2920–2926 (2009)
38. Hong, W.G., Kim, B.H., Lee, S.M., Yu, H.Y., Yun, Y.J., Jun, Y., Lee, J.B., Kim, H.J.: Int. J. Hydrogen Energy **37**, 7594–7599 (2012)
39. Chen, C., Zhang, J., Zhang, B., Duan, H.M.: J. Phys. Chem. C **117**, 4337–4344 (2013)
40. Tylianakis, E., Psofogiannakis, G.M., Froudakis, G.E.: J. Phys. Chem. Lett. **1**, 2459–2464 (2010)
41. Aboutalebi, S.H., Aminorroaya-Yamini, S., Nevirkovets, I., Konstantinov, K., Liu, H.K.: Adv. Energy Mater. **12**, 1439–1446 (2012)
42. Herrera-Alonso, M., Abdala, A.A., McAllister, M.J., Aksay, I.A., Prud'homme, R.K.: Langmuir **23**, 10644–10649 (2007)
43. Burress, J.W., Gadipelli, S., Ford, J., Simmons, J.M., Zhou, W., Yildirim, T.: Angew. Chem. Int. Ed. **49**, 8902–8904 (2010)
44. Srinivas, G., Burress, J.W., Fordab, J., Yildirim, T.: J. Mater. Chem. **21**, 11323–11329 (2011)
45. Chan, Y., Hill, J.M.: Nanotechnology **22**, 305403–305410 (2011)
46. Kim, B.H., Hong, W.G., Moon, H.R., Lee, S.M., Kim, J.M., Kang, S., Jun, Y., Kim, H.J.: Int. J. Hydrogen Energy **37**, 14217–14222 (2012)
47. Kumar, R., Suresh, V.M., Maji, T.K., Rao, C.N.R.: Chem. Commun. **50**, 2015–2017 (2014)
48. Petit, C., Bandosz, T.J.: Adv. Mater. **21**, 4753–4757 (2009)
49. Petit, C., Burress, J., Bandosz, T.J.: Carbon **49**, 563–572 (2011)
50. Liu, S., Sun, L.X., Xu, F., Zhang, J., Jiao, C.L., Li, F., Li, Z.B., Wang, S., Wang, Z.Q., Jiang, X., Zhou, H.Y., Yang, L.N., Schick, C.: Energy Environ. Sci. **6**, 818–823 (2013)
51. Kim, T.K., Cheon, J.Y., Yoo, K., Kim, J.W., Hyun, S.M., Shin, H.S., Joo, S.H., Moon, H.R.: J. Mater. Chem. A **1**, 8432–8437 (2013)
52. Zhou, H., Liu, X.Q., Zhang, J., Yan, X.F., Liu, Y.J., Yuan, A.H.: Int. J. Hydrogen Energy **39**, 2160–2167 (2014)
53. Ott, L.S., Finke, R.G.: Coord. Chem. Rev. **251**, 1075–1100 (2007)
54. Xi, P.X., Chen, F.J., Xie, G.J., Ma, C., Liu, H.Y., Shao, C.W., Wang, J., Xu, Z.H., Xu, X.M., Zeng, Z.Z.: Nanoscale **4**, 5597–5601 (2012)
55. Zhao, W.F., Wang, M., Wu, F.R., Wu, H., Wang, L., Chen, G.H.: J. Mater. Chem. **20**, 5817–5819 (2010)
56. Kılıç, B., Şencanlı, S., Metin, Ö.: J. Mol. Catal. A: Chem. **361–362**, 104–110 (2012)
57. Cao, N., Luo, W., Cheng, G.Z.: Int. J. Hydrogen Energy **38**, 11964–11972 (2013)
58. Yang, L., Luo, W., Cheng, G.Z.: ACS Appl. Mater. Interfaces **5**, 8231–8240 (2013)
59. Cao, N., Su, J., Luo, W., Cheng, G.Z.: Int. J. Hydrogen Energy **39**, 426–435 (2014)
60. Wang, X., Liu, D.P., Song, S.Y., Zhang, H.J.: Chem. Eur. J. **19**, 8082–8086 (2013)
61. Wang, J., Qin, Y.L., Liu, X., Zhang, X.B.: J. Mater. Chem. **22**, 12468–12470 (2012)
62. Zhou, X.H., Chen, Z.X., Yan, D.H., Lu, H.B.: J. Mater. Chem. **22**, 13506–13516 (2012)
63. Yang, Y.W., Lu, Z.H., Hu, Y.J., Zhang, Z.J., Shi, W.W., Chen, X.S., Wang, T.T.: RSC Adv. **4**, 13749–13752 (2014)

64. Yang, Y.W., Zhang, F., Wang, H.L., Yao, Q.L., Chen, X.S., Lu, Z.H. J. Nanomater. Article ID: 294350 (2014)
65. Tang, Z.W., Chen, H., Chen, X.W., Wu, L.M., Yu, X.B.: J. Am. Chem. Soc. **134**, 5464–5467 (2012)
66. Li, F., Gao, J.F., Zhang, J., Xu, F., Zhao, J.J., Sun, L.X.: J. Mater. Chem. A **1**, 8016–8022 (2013)
67. Zhang, J.X., Cao, H.Q., Tang, X.L., Fan, W.F., Peng, G.C., Qu, M.Z.: J. Power Sources **41**, 619–626 (2013)
68. Wang, S.C., Yang, J., Zhou, X.Y., Li, J.: Int. J. Electrochem. Sci. **8**, 9692–9703 (2013)
69. Wan, D.Y., Yang, C.Y., Lin, T.Q., Tang, Y.F., Zhou, M., Zhong, Y.J., Huang, F.Q., Lin, J.H.: ASC Nano **6**, 9068–9078 (2012)
70. Kuo, S.L., Liu, W.R., Kuo, C.P., Wu, N.L., Wu, H.C.: J. Power Sources **244**, 552–556 (2013)
71. Mukherjee, R., Thomas, A.V., Krishnamurthy, A., Koratkar, N.: ASC Nano **6**, 7867–7878 (2012)
72. Wang, H.L., Cui, L F., Yang, Y., Casalongue, H.S., Robinson, J.T., Liang, Y.Y., Cui, Y., Dai, H.J.: J. Am. Chem. Soc. **132**, 13978–13980 (2010)
73. Zhu, J.X., Zhu, T., Zhou, X.Z., Zhang, Y.Y., Lou, X.W., Chen, X.D., Zhang, H., Hng, H.H., Yan, Q.Y.: Nanoscale **3**, 1084–1089 (2011)
74. Gao, T., Huang, K., Qi, X., Li, H.X., Yang, L.W., Zhong, J.X.: Ceram. Int. **140**, 6891–6897 (2014)
75. Ni, H.F., Song, W.L., Fan, L.Z.: Electrochem. Comm. **40**, 1–4 (2014)
76. Zou, F., Hu, X.L., Qie, L., Jiang, Y., Xiong, X.Q., Qiao, Y., Huang, Y.H.: Nanoscale **6**, 924–930 (2014)
77. Qiu, B., Zhao, X.Y., Xia, D.G.: J. Alloy. Compd. **579**, 372–376 (2013)
78. Fei, L., Lin, Q.L., Yuan, B., Chen, G., Xie, P., Li, Y.L., Xu, Y., Deng, S.G., Smirnov, S., Luo, H.M.: ACS Appl. Mater. Interfaces **5**, 5330–5335 (2013)
79. Chen, D.Z., Quan, H.Y., Luo, X.B., Luo, S.L.: Scripta Mater. **76**, 1–4 (2014)
80. Chan, C.K., Peng, H., Liu, G., McIlwrath, K., Zhang, X.F., Huggins, R.A., Cui, Y.: Nat. Nanotechnol. **3**, 31–35 (2008)
81. Kasavajjula, U., Wang, C., Appleby, A.J.: J. Power Sources **163**, 1003–1039 (2007)
82. Guo, Z.Y., Wang, J., Wang, F., Zhou, D.D., Xia, Y.Y., Wang, Y.G.: Adv. Funct. Mater. **23**, 4840–4846 (2013)
83. Hurley, P.T., Chen, J.H.: Adv. Mater. **26**, 758–764 (2014)
84. Ren, J.G., Wang, C.D., Wu, Q.H., Liu, X., Yang, Y., He, L.F., Zhang, W.J.: Nanoscale **6**, 3353–3360 (2014)
85. Stournara, M.E., Shenoy, V.B.: J. Power Sources **196**, 5697–5703 (2011)
86. Wang, D.W., Sun, C.H., Zhou, G.M., Li, F., Wen, L., Donose, B.C., Lu, G.Q., Cheng, H.M., Gentle, I.R.: J. Mater. Chem. A **1**, 3607–3612 (2013)
87. Ha, S.H., Jeong, Y.S., Lee, Y.J.: ACS Appl. Mater. Interfaces **5**, 12295–12303 (2013)
88. Tu, F.Y., Liu, S.Q., Wu, T.H., Jin, G.H., Pan, C.Y.: Powder Technol. **253**, 580–583 (2014)
89. Pei, B., Jiang, Z.Q., Zhang, W.X., Yang, Z.H., Manthiram, A.: J. Power Sources **239**, 475–482 (2013)
90. Wu, K.L., Yang, J.P.: Mater. Res. Bull. **48**, 435–439 (2013)
91. Rui, X.H., Zhu, J.X., Sim, D.H., Xu, C., Zeng, Y., Hng, H.H., Lim, T.M., Yan, Q.Y.: Nanoscale **3**, 4752–4758 (2011)
92. Zhao, H.B., Pan, L.Y., Xing, S.Y., Luo, J., Xu, J.Q.: J. Power Sources **222**, 21–31 (2013)
93. Pham-Cong, D., Ahn, K., Hong, S.W., Jeong, S.Y., Choi, J.H., Doh, C.H., Jin, J.S., Jeong, E.D., Cho, C.R.: Curr. Appl. Phys. **14**, 215–221 (2014)
94. Chen, D.Z., Quan, H.Y., Luo, S.L., Luo, X.B., Deng, F., Jiang, H.L.: Phys. E **56**, 231–237 (2014)
95. Yamin, H., Peled, E.: J. Power Sources **9**, 281–287 (1983)
96. Elazari, R., Salitra, G., Talyosef, Y., Grinblat, J., Scordilis-Kelley, C., Xiao, A., Affinito, J., Aurbach, D.: J. Electrochem. Soc. **157**, A1131–A1138 (2010)
97. Cheon, S.E., Ko, K.S., Cho, J.H., Kim, S.W., Chin, E.Y., Kim, H.T.: J. Electrochem. Soc. **150**, A800–A805 (2003)

98. Jeon, B.H., Yeon, J.H., Kim, K.M., Chung, I.J.: J. Power Sources **109**, 89–97 (2002)
99. Mikhaylik, Y.V., Akridge, J.R.: J. Electrochem. Soc. **151**, A1969–A1976 (2004)
100. Kim, J., Cote, L.J., Huang, J.: Acc. Chem. Res. **45**, 1356–1364 (2012)
101. Lightcap, I.V., Kamat, P.V.: Acc. Chem. Res. **46**, 2235–2243 (2013)
102. Ji, L.W., Rao, M.M., Zheng, H.M., Zhang, L., Li, Y.C., Duan, W.H., Guo, J.H., Cairns, E.J., Zhang, Y.G.: J. Am. Chem. Soc. **133**, 1852–18525 (2011)
103. Zhang, L., Ji, L.W., Glans, P.A., Zhang, Y.G., Zhu, J.F., Guo, J.H.: Phys. Chem. Chem. Phys. **14**, 13670–13675 (2012)
104. Wang, H.L., Yang, Y., Liang, Y.Y., Robinson, J.T., Li, Y.G., Jackson, A., Cui, Y., Dai, H.J.: Nano Lett. **11**, 2644–2647 (2011)
105. Xiao, M., Huang, M., Zeng, S.S., Han, D.M., Wang, S.J., Sun, L.Y., Meng, Y.Z.: RSC Adv. **3**, 4914–4916 (2013)
106. Kim, J.W., Ocon, J.D., Park, D.W., Lee, J.: J. Energy Chem. **22**, 336–340 (2013)
107. Song, M.K., Zhang, Y.G., Cairns, E.J.: Nano Lett. **13**, 5891–5899 (2013)
108. Zhou, W.D., Chen, H., Yu, Y.C., Wang, D.L., Cui, Z.M., DiSalvo, F.J., Abruña, H.D.: ASC Nano **7**, 8801–8808 (2013)
109. Seh, Z.W., Wang, H.T., Liu, N., Zheng, G.Y., Li, W.Y., Yao, H.B., Cui, Y.: Chem. Sci. **5**, 1396–1400 (2014)
110. Yang, Y., Zheng, G., Misra, S., Nelson, J., Toney, M.F., Cui, Y.: J. Am. Chem. Soc. **134**, 15387–15394 (2012)
111. Zhang, F.F., Zhang, X.B., Dong, Y.H., Wang, L.M.: J. Mater. Chem. **22**, 11452–11454 (2012)
112. Han, K., Shen, J.M., Hayner, C.M., Ye, H.Q., Kung, M.C., Kung, H.H.: J. Power Sources **251**, 331–337 (2014)
113. Gao, X.F., Li, J.Y., Guan, D.S., Yuan, C.: ACS Appl. Mater. Interfaces **6**, 4154–4159 (2014)
114. Lu, S.T., Chen, Y., Wu, X.H., Wang, Z.D., Li, Y.: Sci. Rep. **4**, 4629 (2014)
115. Li, N.W., Zheng, M.B., Lu, H.L., Hu, Z.B., Shen, C.F., Chang, X.F., Ji, G.B., Cao, J.M., Shi, Y.: Chem. Commun. **48**, 4106–4108 (2012)
116. Wang, X.F., Wang, Z.X., Chen, L.Q.: J. Power Sources **242**, 65–69 (2013)
117. Li, N.W., Zheng, M.B., Lu, H.L., Hu, Z.B., Shen, C.F., Chang, X.F., Ji, G.B., Cao, J.M., Shi, Y.: Chem. Commun. **48**, 4106–4108 (2012)
118. Zhou, X.Y., Xie, J., Yang, J., Zou, Y.L., Tang, J.J., Wang, S.C., Ma, L.L., Liao, Q.C.: J. Power Sources **243**, 993–1000 (2013)
119. Chen, R.J., Zhao, T., Lu, J., Wu, F., Li, L., Chen, J.Z., Tan, G.Q., Ye, Y.S., Amine, K.: Nano Lett. **13**, 4642–4649 (2013)
120. Xie, J., Yang, J., Zhou, X.Y., Zou, Y.L., Tang, J.J., Wang, S.C., Chen, F.: J. Power Sources **253**, 55–63 (2014)
121. Lu, L.Q., Lu, L.J., Wang, Y.: J. Mater. Chem. A **1**, 9173–9181 (2013)
122. Shao, Y.Y., Zhang, S., Engelhard, M.H., Li, G.S., Shao, G.C., Wang, Y., Liu, J., Aksay, I.A., Lin, Y.H.: J. Mater. Chem. **20**, 7491–7496 (2010)
123. Wang, X.W., Zhang, Z.A., Qu, Y.H., Lai, Y.Q., Li, J.: J. Power Sources **256**, 361–368 (2014)
124. Beattie, S.D., Manolescu, D.M., Blair, S.L.: J. Electrochem. Soc. **156**, A44–A47 (2009)
125. Girishkumar, G., McCloskey, B., Luntz, A.C., Swanson, S., Wilcke, W.: J. Phys. Chem. Lett. **1**, 2193–2203 (2010)
126. Christensen, J., Albertus, P., Sanchez-Carrera, R.S., Lohmann, T., Kozinsky, B., Liedtke, R., Ahmed, J., Kojic, A.: J. Electrochem. Soc. **159**, R1–R30 (2012)
127. Kraytsberg, A., Ein-Eli, Y.: J. Power Sources **196**, 886–893 (2011)
128. Itkis, D.M., Semenenko, D.A., Kataev, E.Y., Belova, A.I., Neudachina, V.S., Sirotina, A.P., Hävecker, M., Teschner, D., Knop-Gericke, A., Dudin, P., Barinov, A., Goodilin, E.A., Shao-Horn, Y., Yashina, L.V.: Nano Lett. **13**, 4697–4701 (2013)
129. Yang, X.H., He, P., Xia, Y.Y.: Electrochem. Commun. **11**, 1127–1130 (2009)
130. Mirzaeian, M., Hall, P.J.: Electrochim. Acta **54**, 7444–7451 (2009)
131. Xiao, J., Mei, D.H., Li, X.L., Xu, W., Wang, D.Y., Graff, G.L., Bennett, W.D., Nie, Z.M., Saraf, L.V., Aksay, I.A., Liu, J., Zhang, J.G.: Nano Lett. **11**, 5071–5078 (2011)

132. Wang, Z.L., Xu, D., Xu, J.J., Zhang, L.L., Zhang, X.B.: Adv. Funct. Mater. **22**, 3699–3705 (2012)
133. Lai, L.F., Potts, J.R., Zhan, D., Wang, L., Poh, C.K., Tang, C.H., Gong, H., Shen, Z.X., Lin, J.Y., Ruoff, R.S.: Energy Environ. Sci. **5**, 7936–7942 (2012)
134. Jung, H.G., Jeong, Y.S., Park, J.B., Sun, Y.K., Scrosati, B., Lee, Y.J.: ACS Nano **7**, 3532–3539 (2013)
135. Liang, Y.Y., Li, Y.G., Wang, H.L., Zhou, J.G., Wang, J., Regier, T., Dai, H.J.: Nat. Mater. **10**, 780–786 (2011)
136. Li, Q., Xu, P., Zhang, B., Tsai, H., Wang, J., Wang, H.L., Wu, G.: Chem. Commun. **49**, 10838–10840 (2013)
137. Zhu, Y.W., Murali, S., Cai, W.W., Li, X.S., Suk, J.W., Potts, J.R., Ruoff, R.S.: Adv. Mater. **22**, 5226 (2010)
138. Zabihinpour, M., Ghenaatian, H.R.: Synth. Met. **175**, 62–67 (2013)
139. Shulga, Y.M., Baskakov, S.A., Smirnov, V.A., Shulga, N.Y., Belay, K.G., Gutsev, G.L.: J. Power Sources **245**, 33–36 (2014)
140. Lin, Z.Y., Liu, Y., Yao, Y.G., Hildreth, O.J., Li, Z., Moon, K., Wong, C.P.: J. Phys. Chem. C **115**, 7120–7125 (2011)
141. Zhang, D.C., Zhang, X., Chen, Y., Wang, C.H., Ma, Y.W.: Electrochim. Acta **69**, 364–370 (2012)
142. Bai, Y.C., Rakhi, R.B., Chen, W., Alshareef, H.N.: J. Power Sources **233**, 313–319 (2013)
143. Zhao, B., Liu, P., Jiang, Y., Pan, D.Y., Tao, H.H., Song, J.S., Fang, T., Xu, W.W.: J. Power Sources **198**, 423–427 (2012)
144. Zhang, W., Zhang, Y.X., Tian, Y., Yang, Z.Y., Xiao, Q.Q., Guo, X., Jing, L., Zhao, Y.F., Yan, Y.M., Feng, J.S., Sun, K.N.: ACS Appl. Mater. Interfaces **6**, 2248–2254 (2014)
145. Han, J.W., Zhang, L.L., Lee, S.J., Oh, J.H., Lee, K.S., Potts, J.R., Ji, J.Y., Zhao, X., Ruoff, R.S., Park, S.J.: ACS Nano **7**, 19–26 (2013)
146. Gopalakrishnan, K., Moses, K., Govindaraj, A., Rao, C.N.R.: Solid State Commun. **175–176**, 43–50 (2013)
147. Nolan, H., Mendoza-Sanchez, B., Kumar, N.A., McEvoy, N., O'Brien, S., Nicolosi, V., Duesberg, G.S.: Phys. Chem. Chem. Phys. **16**, 2280–2284 (2014)
148. Yang, J., Jo, M.R., Kang, M., Huh, Y.S., Jung, H., Kang, Y.M.: Carbon **73**, 106–113 (2014)
149. Lee, J.H., Park, N., Kim, B.G., Jung, D.S., Im, K., Hur, J., Choi, J.W.: ACS Nano **7**, 9366–9374 (2013)
150. Yoon, Y., Lee, K., Baik, C., Yoo, H., Min, M., Park, Y., Lee, S.M., Lee, H.: Adv. Mater. **25**, 4437–4444 (2013)
151. Yan, J., Wang, Q., Wei, T., Jiang, L.L., Zhang, M.L., Jing, X.Y., Fan, Z.J.: ACS Nano **8**, 4720–4729 (2014)
152. Maiti, U.N., Lim, J., Lee, K.E., Lee, W.J., Kim, S.O.: Adv. Mater. **26**, 615–619 (2014)
153. Acik, M., Mattevi, C., Gong, C., Lee, G., Cho, K., Chhowalla, M., Chabal, Y.J.: ACS Nano **4**, 5861–5868 (2010)
154. Zhu, Y.W., Murali, S., Cai, W.W., Li, X.S., Suk, J.W., Potts, J.R., Ruoff, R.S.: Adv. Mater. **22**, 3906–3924 (2010)
155. Chen, W., Yan, L.: Adv. Mater. **24**, 6229–6233 (2012)
156. Luo, J.Y., Jang, H.D., Huang, J.X.: ACS Nano **7**, 1464–1471 (2013)
157. Dong, X.C., Xing, G.C., Chan-Park, M.B., Shi, W.H., Xiao, N., Wang, J., Yan, Q.Y., Sum, T.C., Huang, W., Chen, P.: Carbon **49**, 5071–5078 (2011)
158. Lin, L.Y., Yeh, M.H., Tsai, J.T., Huang, Y.H., Sun, C.L., Ho, K.C.: J. Mater. Chem. A **1**, 11237–11245 (2013)
159. Zeng, F.Y., Kuang, Y.F., Zhang, N.S., Huang, Z.Y., Pan, Y., Hou, Z.H., Zhou, H.H., An, C.L., Schmidt, O.G.: J. Power Sources **247**, 396–401 (2014)
160. Beidaghi, M., Wang, C.L.: Adv. Funct. Mater. **22**, 4501–4510 (2012)
161. Wang, S.Y., Dryfe, R.A.W.: J. Mater. Chem. A **1**, 5279–5283 (2013)
162. Chartarrayawadee, W., Moulton, S.E., Too, C.O., Kim, B.C., Yepuri, R., Romeo, T., Gordon Wallace, G.: J. Appl. Electrochem. **43**, 865–877 (2013)

163. Liu, J.Z., Galpaya, D., Notarianni, M., Yan, C., Motta, N.: Appl. Phys. Lett. **103**, 063108–063111 (2013)
164. Cao, Y.C., Zhu, M., Li, P.X., Zhang, R.J., Li, X.M., Gong, Q.M., Wang, K.L., Zhong, M.L., Wu, D.H., Lin, F., Zhu, H.W.: Phys. Chem. Chem. Phys. **15**, 19550–19556 (2013)
165. Tian, L.L., Meziani, M.J., Lu, F., Kong, C.Y., Cao, L., Thorne, T.J., Sun, Y.P.: ACS Appl. Mater. Interfaces **11**, 3217–3222 (2010)
166. Zhang, C., Ren, L.L., Wang, X.Y., Liu, T.X.: J. Phys. Chem. C **114**, 11435–11440 (2010)
167. Tung, V.C., Huang, J.H., Tevis, I., Kim, F., Kim, J., Chu, C.W., Stupp, S.I., Huang, J.X.: J. Am. Chem. Soc. **133**, 4940–4947 (2011)
168. Cheng, Q., Tang, J., Ma, J., Zhang, H., Shinya, N., Qin, L.C.: Phys. Chem. Chem. Phys. **13**, 17615–17624 (2011)
169. Wang, Y., Wu, Y., Huang, Y., Zhang, F., Yang, X., Ma, Y., Chen, Y.: J. Phys. Chem. C **115**, 23192–23197 (2011)
170. Yang, S.Y., Chang, K.H., Tien, H.W., Lee, Y.F., Li, S.M., Wang, Y.S., Wang, J.Y., Ma, C.C.M., Hu, C.C.: J. Mater. Chem. **21**, 2374–2380 (2011)
171. Liu, R., Lee, S.B.: J. Am. Chem. Soc. **130**, 2942–2943 (2008)
172. Sun, Y.Q., Wu, Q.O., Shi, G.Q.: Energy Environ. Sci. **4**, 1113–1132 (2011)
173. Simon, P., Gogotsi, Y.: Nat. Mater. **7**, 845–854 (2008)
174. Chen, S., Zhu, J.W., Wu, X.D., Han, Q.F., Wang, X.: ACS Nano **4**, 2822–2830 (2010)
175. Gund, G.S., Dubal, D.P., Patil, B.H., Shinde, S.S., Lokhande, C.D.: Electrochim. Acta **92**, 205–215 (2013)
176. Antiohos, D., Pingmuang, K., Romano, M.S., Beirne, S., Romeo, T., Aitchison, P., Minett, A., Wallace, G., Phanichphant, S., Chen, J.: Electrochim. Acta **101**, 99–108 (2013)
177. Li, Y., Zhao, N.Q., Shi, C.S., Liu, E.Z., He, C.N.: J. Phys. Chem. C **116**, 25226–25232 (2012)
178. Kim, M., Hwang, Y., Kim, J.: J. Power Sources **239**, 225–233 (2013)
179. Sumboja, A., Foo, C.Y., Wang, X., Lee, P.S.: Adv. Mater. **25**, 2809–2815 (2013)
180. Zhou, W.W., Liu, J.P., Chen, T., Tan, K.S., Jia, X.T., Luo, Z.Q., Cong, C.X., Yang, H.P., Li, C.M., Yu, T.: Phys. Chem. Chem. Phys. **13**, 14462–14465 (2011)
181. Zhang, D.H., Zou, W.B.: Curr. Appl. Phys. **13**, 1796–1800 (2013)
182. Yuan, C.Z., Zhang, L.H., Hou, L.R., Pang, G., Oh, W.C.: RSC Adv. **4**, 14408–14413 (2014)
183. Shi, W.H., Zhu, J.X., Sim, D.H., Tay, Y.Y., Lu, Z.Y., Zhang, X.J., Sharma, Y., Srinivasan, M., Zhang, H., Hng, H.H., Yan, Q.Y.: J. Mater. Chem. **21**, 3422–3427 (2011)
184. Cheng, J.P., Shou, Q.L., Wu, J.S., Liu, F., Dravid, V.P., Zhang, X.B.J.: Electroanal. Chem. **698**, 1–8 (2013)
185. Khoh, W.H., Hong, J.D.: Colloids Surf., A **436**, 104–112 (2013)
186. He, G.Y., Wang, L., Chen, H.Q., Sun, X.Q., Wang, X.: Mater. Lett. **98**, 164–167 (2013)
187. Li, W., Bu, Y.F., Jin, H.L., Wang, J., Zhang, W.M., Wang, S., Wang, J.C.: Energy Fuels **27**, 6304–6310 (2013)
188. Lee, G., Cheng, Y.W., Varanasi, C.V., Liu, J.: J. Phys. Chem. C **118**, 2281–2286 (2014)
189. Prakash, A., Bahadur, D.: ACS Appl. Mater. Interfaces **6**, 1394–1405 (2014)
190. Liu, Y., Ying, Y.L., Mao, Y.Y., Gu, L., Wang, Y.W., Peng, X.S.: Nanoscale **5**, 9134–9140 (2013)
191. Xiang, C.C., Li, M., Zhi, M.J., Manivannan, A., Wu, N.Q.: J. Mater. Chem. **22**, 19161–19167 (2012)
192. Wang, H.L., Hao, Q.L., Yang, X.J., Lu, L.D., Wang, X.: Electrochem. Commun. **11**, 1158–1161 (2009)
193. Xu, J.J., Wang, K., Zu, S.Z., Han, B.H., Wei, Z.X.: ACS Nano **4**, 5019–5026 (2010)
194. Wang, H.L., Hao, Q.L., Yang, X.J., Lu, L.D., Wang, X.: Appl. Mater. Interfaces **2**, 821–828 (2010)
195. Liu, Y., Deng, R.J., Wang, Z., Liu, H.T.: J. Mater. Chem. **22**, 13619–13624 (2012)
196. Kumar, N.A., Choi, H.J., Shin, Y.R., Chang, D.W., Dai, L.M., Baek, J.B.: ACS Nano **6**, 1715–1723 (2012)
197. Lai, L.F., Yang, H.P., Wang, L., Teh, B.K., Zhong, J.Q., Chou, H., Chen, L.W., Chen, W., Shen, Z.X., Ruoff, R.S., Lin, J.Y.: ACS Nano **6**, 5941–5951 (2012)

198. Wang, L., Ye, Y.J., Lu, X.P., Wen, Z.B., Li, Z., Hou, H.Q., Song, Y.H.: Sci. Rep. **3**, 3568–3576 (2013)
199. Kim, M., Lee, C., Jang, J.: Adv. Funct. Mater. **24**, 2489–2499 (2014)
200. Li, J., Xie, H.Q.: Mater. Lett. **78**, 106–109 (2012)
201. Zhu, C.Z., Zhai, J.F., Wen, D., Dong, S.J.: J. Mater. Chem. **22**, 6300–6306 (2012)
202. Chen, Z., Yu, D.S., Xiong, W., Liu, P.P., Liu, Y., Dai, L.M.: Langmuir **30**, 3567–3571 (2014)
203. Liu, Y., Zhang, Y., Ma, G.H., Wang, Z., Liu, K.Y., Liu, H.T.: Electrochim. Acta **88**, 519–525 (2013)
204. Park, H.S., Lee, M.H., Hwang, R.Y., Park, O.K., Jo, K., Lee, T., Kim, B.S., Song, H.K.: Nano Energy **3**, 1–9 (2014)
205. Wee, B.H., Hong, J.D.: Langmuir **30**, 5267–5275 (2014)
206. Yang, X., Zhang, F., Zhang, L., Zhang, T.F., Huang, Y., Chen, Y.S.: Adv. Funct. Mater. **23**, 3353–3360 (2013)
207. Yang, X., Zhang, L., Zhang, F., Zhang, T.F., Huang, Y., Chen, Y.S.: Carbon **72**, 381–386 (2014)
208. Zhang, K., Ang, B.T., Zhang, L.L., Zhao, X.S., Wu, J.S.: J. Mater. Chem. **21**, 2663–2670 (2011)
209. Cui, Y., Cheng, Q.Y., Wu, H.P., Wei, Z.X., Han, B.H.: Nanoscale **5**, 8367–8374 (2013)
210. Zhang, J.T., Zhao, X.S.: J. Phys. Chem. C **116**, 5420–5426 (2012)
211. Wang, Y.S., Li, S.M., Hsiao, S.T., Liao, W.H., Chen, P.H., Yang, S.Y., Tien, H.W., Ma, C.C.M., Hu, C.C.: Carbon **73**, 87–98 (2014)
212. Trigueiro, J.P.C., Lavall, R.L., Silva, G.G.: J. Power Sources **256**, 264–273 (2014)
213. Prasad, K.R., Munichandraiah, N.: J. Power Sources **112**, 443–451 (2002)
214. Palaniappan, S., Devi, S.L.: J. Appl. Polym. Sci. **107**, 1887–1892 (2008)
215. Li, C., Bai, H., Shi, G.: Chem. Soc. Rev. **38**, 2397–2409 (2009)
216. Ratha, S., Sekhar Rout, C.S.: ACS Appl. Mater. Interfaces **5**, 11427–11433 (2013)
217. Wang, A.M., Wang, H.L., Zhang, S.Y., Mao, C.J., Song, J.M., Niu, H.L., Jin, B.K., Tian, Y.P.: Appl. Surf. Sci. **282**, 704–708 (2013)
218. Ciszewski, M., Mianowski, A., Nawrat, G., Szatkowski, P.: ISRN Electrochem. Article ID: 826832 (2014)
219. Ouyang, W.Z., Sun, J.H., Memon, J., Wang, C., Geng, J.X., Huang, Y.: Carbon **62**, 501–509 (2013)
220. Han, X., Liu, S.J., Yuan, Y., Wang, Y., Hu, L.J.: J. Alloy. Compd. **543**, 200–205 (2012)
221. Xu, D.D., Xu, Q., Wang, K.X., Chen, J., Chen, Z.M.: ACS Appl. Mater. Interfaces **6**, 200–209 (2014)
222. Yang, C.Y., Shen, J.L., Wang, C.Y., Fei, H.J., Bao, H., Wang, G.C.: J. Mater. Chem. A **2**, 1458–1464 (2014)
223. Su, H.F., Wang, T., Zhang, S.Y., Song, J.M., Mao, C.J., Niu, H.L., Jin, B.K., Wu, J.Y., Tian, Y.P.: Solid State Sci. **14**, 677–681 (2012)
224. Mu, B., Zhang, W.B., Shao, S.J., Wang, A.Q.: Phys. Chem. Chem. Phys. **16**, 7872–7880 (2014)
225. Yuan, B.Q., Xu, C.Y., Deng, D.H., Xing, Y., Liu, L., Pang, H., Zhang, D.J.: Electrochim. Acta **88**, 708–712 (2013)
226. Sawangphruk, M., Srimuk, P., Chiochan, P., Krittayavathananon, A., Luanwuthi, S., Limtrakul, J.: Carbon **60**, 109–116 (2013)
227. Yan, L., Li, R.Y., Li, Z.J., Liu, J.K., Fang, Y.J., Wang, G.L., Gu, Z.G.: Electrochim. Acta **95**, 146–154 (2013)
228. Yang, W.L., Gao, Z., Wang, J., Ma, J., Zhang, M.L., Liu, L.H.: ACS Appl. Mater. Interfaces **5**, 5443–5454 (2013)
229. Zhang, L.L., Xiong, Z.G., Zhao, X.S.: J. Power Sources **222**, 326–332 (2013)

Chapter 6
Application of GO in Environmental Science

Abstract The industrial and agricultural activities of human induce environmental pollution in air and water. Especially, greenhouse and toxic gases, heavy metal ions and organic species are deteriorating the ecological balance and human health everyday. Owing to the unique surface chemistry and architectures, GO and its composites have versatile applications in environmental protection. Metal-decorated GO can be used for capture and conversion of the harmful gases such as CO_2, CO, NO_2, and NH_3. Another important environmental application of GO is water purification. The photocatalytic activity of GO for the removal of heavy metal ions from polluted water can be significantly enhanced by forming hybrid with metal oxides. Additionally, GO-based composites can be used for photodegradation of organic pollutants in wastewater.

6.1 Air Pollutant Removal

6.1.1 Gas Adsorption and Detection

Among all kinds of air pollutants, carbon dioxide (CO_2) contributes largely to the global climate change, which is in urgent need to be alleviated. To remove CO_2 gases, the aminated graphite oxides have been used for CO_2 adsorption and achieved a high adsorption capacity of 46.55 mg CO_2/g sample [1]. Later, the few-layer GO sheets exhibited superior gas separation characteristics of carbon dioxide/nitrogen, which render the well-interlocked GO membranes suitable for postcombustion CO_2 capture processes, even in a humidified circumstance [2]. Furthermore, the water molecules were found to play a significant role in the intercalation of CO_2 into GO layers [3]. Once the water vanishes, the CO_2 intercalated GO structure decomposes and the water incorporation is accompanied with formation of $C^{16}O^{18}O$, as shown in Fig. 6.1. The key role of water in CO_2 intercalation between GO interlayers has been reproved by Kim and coworkers [4]. They observed a remarkable enhancement of CO_2 adsorption in water-swelled GO

© The Author(s) 2015
J. Zhao et al., *Graphene Oxide: Physics and Applications*,
SpringerBriefs in Physics, DOI 10.1007/978-3-662-44829-8_6

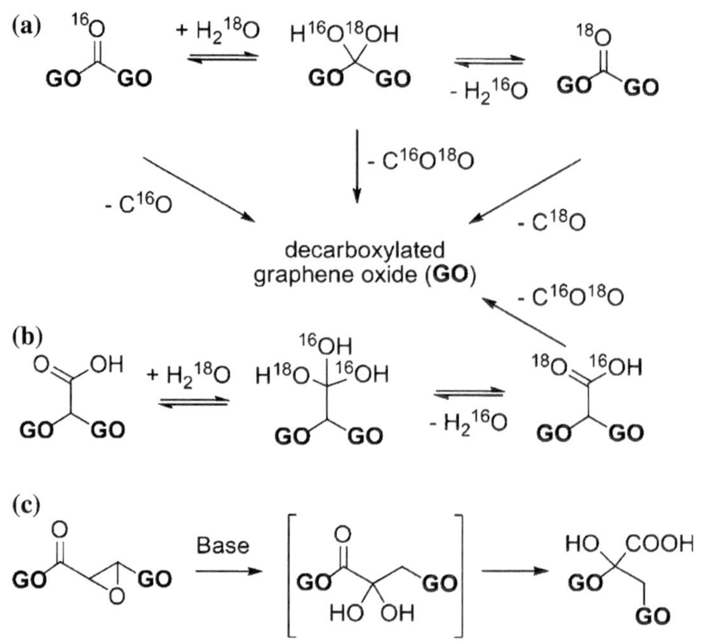

Fig. 6.1 a Exchange of water and hydrate formation leading to CO and CO_2; **b** exchange of water in carboxylic acids; **c** rearrangement reaction of α-epoxy ketones leading to carboxylic acids. Reprinted with permission from Ref. [3]. Copyright (2012) American Chemical Society

compared to the dried one. Moreover, the water circumstance also facilitates the intercalation of other gases such as CH_4, N_2, and H_2. Their MD simulation indicated that the intercalated water molecules and the functional groups on the GO plane contribute to the increase of CO_2 adsorption. A very recent DFT calculation indicates that the intercalation of water molecules can influence the GO interlayer spacing, and therefore affects the CO_2 migration within the GO layers by the repulsive interactions between CO_2 and oxygen-containing groups attached on the GO sheets [5].

To improve the adsorption capacity of CO_2, various GO composites have been explored [6–11]. The GO-layered double oxides (LDHs) show great potential in CO_2 capture [6]. Compared to the MWCNTs-LDHs [12], less amount of GO is needed in GO-LDHs to achieve the highest CO_2 adsorption capacity of 62 % [6]. However, further increase of GO loading cannot simply improve the CO_2 adsorption due to the poor network forming ability. In addition, the layered double oxides supported by GO can be used for CO_2 capture [7]. Despite that the CO_2 capacity gradually decreases after the first cycle, the presence of GO is able to reduce the loss of adsorption capacity, and helps maintain the heterogeneity of the basic sites on adsorbents for a longer number of temperature-swing cycles.

Polymer-decorated GO also shows positive effects, especially in enhancing the surface area and improving the cycle stability for CO_2 capture [8–10]. The RGO/PANI composite can provide a uniformly N-doped porous graphene material with a maximum BET (Brunauer, Emmet and Teller) surface area of 1,336 m^2 g^{-1} [8]. This N-doped graphene shows a highly reversible maximum CO_2 storage capacity of 2.7 mmol g^{-1} at 298 K and 1 atm (5.8 mmol g^{-1} at 273 K and 1 atm) along with a good recycling stability with only an initial reduction of 10 %. A polyindole-RGO (PIG) hybrid with a maximum BET surface area of 936 m^2 g^{-1} also shows superior CO_2 capture capacity of 3.0 mmol g^{-1} at 25 °C and 1 atm [9]. Alhwaige et al. [10] firstly demonstrated the highly dispersion of GO within a biopolymer chitosan and observed synergistic effect for CO_2 capture. The 20 wt% GO loaded hybrid architecture exhibits remarkable increase of CO_2 storage capacity from 1.92 to 4.15 mol kg^{-1} and prolongs the adsorption-desorption cyclic life. Besides, the porous Cu-based MOF was demonstrated to be well-dispersed on GO surface with nanometer scale, which is favored with more active sites for CO_2 adsorption and thus leads to a high CO_2 adsorption capacity of 8.26 mmol g^{-1} at 273 K under 1 atm [11].

Apart from the greenhouse gases, the removal of harmful gases via GO based materials was mainly focused on ammonia [13–21]. Bandosz's group has devoted substantial efforts on the ammonia removal with the graphite oxide based materials [13–19]. They found that the surface chemistry of graphite oxides, especially the oxygen-containing groups and the remained sulfonic groups, and the water molecules play significant role on the ammonia adsorption behaviors [13–16]. Moreover, the interaction mechanism of ammonia with GO varies with the synthesis approaches of GO [14]. Further composition of GO and other materials with affinity to ammonia such as polyoxometalate [17], Al_{13} [18] and MnO_2 [19] exhibits improvement of ammonia accommodation compared to the pure GO.

First-principles calculations indicated that the presence of diverse active defect sites (the hydroxyl and epoxy functional groups and their neighboring carbon atoms) on GO can promote the charge transfer between NH_3 and GO and therefore strengthen the adsorption of NH_3 [20]. In situ IR microspectroscopy experiments integrated with DFT calculations also confirmed the co-contribution of epoxide groups and carbon vacancies to NH_3 dissociation, which leads to a wide variety of adsorbed species with a small net electron-donor effect [21].

Removal of other harmful gases like formaldehyde [22], acetone [23], H_2S [24, 25], SO_2 [26, 29], CO [27], nitrogen oxides [28, 29] via GO and its composites has also been addressed. The silylated graphite oxide containing amino group, exhibits superior ability of formaldehyde accommodation, which prevails the activated carbon material even with the presence of water molecules [22]. As for the extremely dangerous acetone gas, the GO foam shows higher adsorption efficiency than that of RGO foams and other carbon materials [23]. In addition, a small amount of GO integrated with MOF-5 can increase the available Zn sites in MOF-5 and also create pore space with strong dispersive forces for H_2S capture [24]. But the H_2S adsorption capacity is reduced since the MOF structure is distorted upon GO loading. Therefore, the glucose was further introduced to maintain the structural stability, leading to a maximum H_2S uptake of 130.1 mg g^{-1}. In order to understand the interaction mechanism of H_2S and GO, Huang et al. [25] constructed a series of surface tunable GO structures for selective adsorption of H_2O/H_2S. They found that H_2S molecules dissociate on the carbonyl functional groups, and that the H_2O molecules act as the challenging gas, which tend to block the potential active sites for H_2S decomposition. Bandosz et al. also found that the $GO/Zr(OH)_4$ composites can be used for SO_2 capture at ambient conditions via reactions with terminal –OH groups of hydrous zirconia [26].

To obtain atomistic insights into the interaction mechanism between GO and the harmful gases, the first-principles calculations have been performed [27–29]. Zhao's group [27] found that the titanium-decorated GO (Ti–GO) can selectively adsorb the carbon monoxide from gas mixtures with a large binding energy of ~70 kJ/mol due to hybridization between the empty d orbitals of Ti and the occupied p orbitals of CO. Tang et al. [28] revealed that the charge transfers between active defect sites of GO (the hydroxyl and carbonyl functional groups and the carbon atom near these groups) and nitrogen oxides can strengthen the adsorption of NO_x (x = 1, 2, 3), eventually leading to chemisorption of the gas molecules. Very recently, it was argued that the light metals (Li and Al) anchored on GO surface exhibit superior adsorption capability of acidic gases compared to that of Ti [29].

6.1.2 Gas Conversion

Even with the advanced progresses in capturing harmful gases, the ultimate goal is to convert them into some useful energy resource for full utilization, which can simultaneously alleviate the energy and environment crisis. In this regard, GO and its composites have attracted many attentions in the harmful gas conversion.

Firstly, the RGO supported cobalt nanocrystals have been documented as catalyst for the Fischer-Tropsch CO_2 hydrogenation process [30]. The oxygen-containing groups on GO can accelerate the formation process of Co nanocrystals, leading to 3-D reticular structure of the composite. Compared to the pure Co catalyst [31], this 3-D reticular RGO/Co composite exhibits superior catalytic activity for CO_2 conversion. In addition, the synergetic interaction of hydroxyl groups on GO with halide anions has a remarkable catalytic influence on the conversion of CO_2 to propylene oxide (PO) at room temperature and atmospheric pressure [32]. The $GO-Bu_4NBr$ composite shows excellent catalytic performance in the cycloaddition of CO_2 to PO under relatively mild conditions, leading to 96 % yield and 100 % selectivity at relatively mild conditions.

Besides, GO shows excellent photocatalytic activity for CO_2 conversion [33–38]. The incorporation of RGO can dramatically enhance the catalytic activity (nearly six times higher than the optimized Cu_2O and 50 times higher than the Cu_2O/RuO_x junction) and the stability of CuO_2 for CO_2 photoreduction to CO [33]. The synergistic effects of retarded electron–hole recombination, efficient charge transfer, and protective function of RGO accounted for the superior

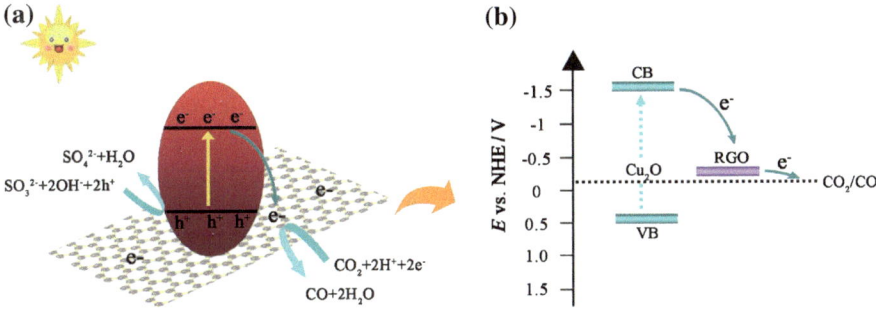

Fig. 6.2 Schematic illustration of the charge transfer in Cu_2O/RGO composites. Reprinted with permission from Ref. [33]. Copyright (2014) The Authors

photocatalytic activity and stability of the Cu_2O/RGO composite (see Fig. 6.2). Moreover, the noble metal (like Pt) modified RGO and TiO_2 nanotubes as cathode catalyst and photo-anode catalysts, respectively, can effectively convert CO_2 into valuable chemicals (CH_3OH, C_2H_5OH, HCOOH and CH_3COOH). The generation rate of liquid products reaches approximate 600 nmol/(h cm^2) and the carbon atom conversion rate reaches 1,130 nmol/(h cm^2), which are much more efficient than those achieved with Pt-modified carbon nanotubes and platinum carbon as cathode catalysts [34].

Despite the superior photocatalytic performance of Pt based composites, the high cost of Pt significantly limits the wide application, which intrigues the exploration of alternative noble metal free catalysts for CO_2 conversion [35–39]. Hence the modified GOs have been developed as the photocatalyst for conversion of CO_2 to methanol [35]. The methanol conversion rate can reach up to 0.172 μmol g cat^{-1} h^{-1} (under 300-W halogen lamp irradiation) by modulating the oxygenated functional groups, much higher than that of pure TiO_2. Moreover, the Co^{II} phthalocyanine complex coated GO (CoPc–GO) has been demonstrated to be an efficient, cost-effective and recyclable photocatalyst for the reduction of carbon dioxide to produce methanol as the main product [36]. Additionally, the RGO-TiO_2 photocatalyst has been developed for converting CO_2 to CH_4 with superior conversion rate (0.135 μmol g cat^{-1} h^{-1}) compared to the graphite oxide and pure anatase [37]. The enhancement of photocatalytic activity is ascribed to the charge anti-recombination by intimate contact of TiO_2 and RGO. Another noble metal free photocatalyst, RGO–CdS nanorod composite, also presents a high CH_4-production rate of 2.51 μmol g cat^{-1} h^{-1}, which exceeds that of pure CdS nanorods by more than ten times and is even better than that of an optimized Pt–CdS nanorod composite photocatalyst under the same reaction conditions [38].

As for the conversion of other harmful gases, it was demonstrated that the porous GO foams not only act as the oxidant but also catalyze the oxidization of SO_2 to SO_3 [39]. Humeres et al. performed thermal reaction followed by plasma treatment to GO and SO_2 and created the oxidized and nonoxidized intermediates intercalated GO, which possess enormous potential in different reactivities with respect to thiolysis and aminolysis, and also selectively adsorb various types of organic moieties [40]. Moreover, ammonia can be converted to the primary amides using rod-like MnO_2/GO hybrid as catalyst [41]. In addition, the colloidal GO exhibits photocatalytic activity for the gas-phase oxidation of benzene by air oxygen [42]. A theoretical study by Jia et al. [43] predicted that the Pd-GO composite exhibits good catalytic performance for CO oxidation. The oxidation reaction proceeds via Langmuir–Hinshelwood mechanism with a two-step route $(CO + O_2 \rightarrow OOCO \rightarrow CO_2 + O)$, followed by the Eley–Rideal mechanism $(CO + O \rightarrow CO_2)$.

6.2 Water Purification

6.2.1 Capture of Pollutants

Pollutants removal from the wastewater via adsorption on the GO based materials has been widely used in the water purification technology because of its high efficiency and low cost. For those heavy metal ions that strongly threaten human body, animal and plants, GO based materials display strong adsorption affinity to remove them from wastewater [44–56].

Among all kinds of heavy metal ions, Cr(VI) ion is considered as one of the most toxic ions due to its high toxicity and bioaccumulation. The ethylenediamine–RGO has been used to remove Cr(VI) from aqueous solutions; however, it takes a very long time (24 h) to achieve 100 mg L^{-1} [44]. Later, the 2, 6-diamino pyridine–RGO composite with the aid of UV-active shows higher Cr(VI) removal ability from acidic water solution (efficiency enhancement by 18 % at a higher pH value) [45]. The presence of an extra pyridinic-nitrogen lone pair facilitates the removal efficiency of excess Cr(VI) [500 mg L^{-1} in 3 h only]. Note that this composite can also partially reduce some Cr(VI) ions, which is advantageous for the Cr(VI) removal. In addition, the conducting polymers decorated GO composites have been demonstrated to be superior sorbents for Cr(VI) adsorption [46, 47]. The hierarchical PANI/GO nanocomposites exhibit excellent water treatment performance with a superb removal capacity of 1,149.4 mg g^{-1} for Cr(VI) [47], which is much higher than other polymer based materials [46, 48–50]. There are also some Cr(VI) ions being partially reduced into Cr(III) species in the PANI/GO composite due to the presence of positive nitrogen groups and the assistance of π electrons on the carbocyclic six-member ring of PANI/GO [47].

GO based composites also exhibit satisfactory adsorption performance for removal of other heavy metal ions [51–56]. Liu et al. [51] reported application of magnetite/GO composite for the adsorbent of Co(II) ions, which exhibits endothermic and spontaneous adsorption process but only achieves an extremely low adsorption capacity of 12.98 mg g^{-1}. Yang et al. [52] found that the Cu^{2+} ions cause GO sheets to be folded and also to form large aggregates, which lead to a huge Cu^{2+}

absorption capacity (around ten times of that of active carbon). Later, the GO aerogels with unidirectional porous structure has been employed for Cu^{2+} removal, which is beneficial for the diffusion of metal ions to the adsorption sites on the GO aerogels [53]. In addition, few-layered GO has been demonstrated with superior adsorption performance of Pb(II) ions owing to the abundant oxygen-containing groups on the surface [54]. The adsorption of Pb(II) ions on GO is a spontaneous and endothermic process, leading to the maximum adsorption capacities (calculated from the Langmuir model) of about 842, 1,150, and 1,850 mg g^{-1} at 293, 313, and 333 K, respectively. Furthermore, the introduction of ethylenediamine triacetic acid (EDTA) to the GO surface induces significant increase of the adsorption capacity for Pb(II) ions [55]. The highest adsorption capacity of EDTA-GO composite for Pb(II) (which varies with pH of the solution) reaches to 479 ± 46 mg g^{-1} at 298 K (± 2), which is 4–5 times higher than that of oxidized carbon nanotubes, and 1–2 times higher than that of pure GO. As for the removal of Hg^{2+} ions, the polypyrrole decorated RGO shows highly selective adsorption capacity of 980 mg g^{-1} and an extremely high desorption capacity of up to 92.3 % [56].

Moreover, the removal of co-existing heavy metal ions from wastewater has been pursued [57–59]. A flower-like TiO$_2$-GO hybrid nanomaterial was utilized for the removal of Zn^{2+}, Cd^{2+} and Pb^{2+} from wastewater [58]. After 6 h and 12 h of hydrothermal treatment at 100 °C, it reaches the maximum adsorption capacities of about 44.8 and 88.9 mg g^{-1} for removing Zn^{2+}, 65.1 and 72.8 mg g^{-1} for removing Cd^{2+}, and 45.0 and 65.6 mg g^{-1} for removing Pb^{2+} (at pH 5.6), respectively, compared to that of colloidal GO under identical condition of about 30.1 (Zn^{2+}), 14.9 (Cd^{2+}), and 35.6 mg g^{-1} (Pb^{2+}). Later, Liu and coworkers [59] reported the removal of Au(III), Pd(II), and Pt(IV) ions via GO itself. The adsorption kinetics of Au(III), Pd(II), and Pt(IV) onto GO follows a pseudosecond-order kinetic model, and eventually the maximum adsorption capacities of 108.342, 80.775, and 71.378 mg g^{-1}, are achieved respectively.

Another kind of harmful water pollutants is organic dyes, which has been widely used in various industrial fields, such as coating, papermaking and textiles. One effective way to removal them is the direct adsorption [60]. To date, GO based materials have made great progresses for the methylene blue (MB) removal [61–65]. Zhang et al. [61] reported that the GO itself is an excellent absorbent for MB removal, with a maximum absorption capacity of 714 mg g^{-1} and the removal efficiency of more than 99 % at initial MB concentrations lower than 250 mg/L. The main strength of absorption is mainly ascribed to the electrostatic interaction and partially contributed by the π-π stacking interaction. Chen et al. [62] also demonstrated that the RGO has strong affinity for MB adsorption. The RGO-N shows a maximum adsorption capacity of 159 mg g^{-1}, which is attributed to the large conjugate structure arising from the interaction between C=O or C=N double bonds and C–C/C=C bonds. Later, He et al. [63] found that expansion of lamellar spacing, maintenance of lamellar structure, and negatively charged oxygen-containing groups are favorable for the MB adsorption. For the GO/RGO composites, it has been demonstrated that the introduction of Mg(OH)$_2$ nanoplates can assist RGO to maintain a high specific surface area and form the

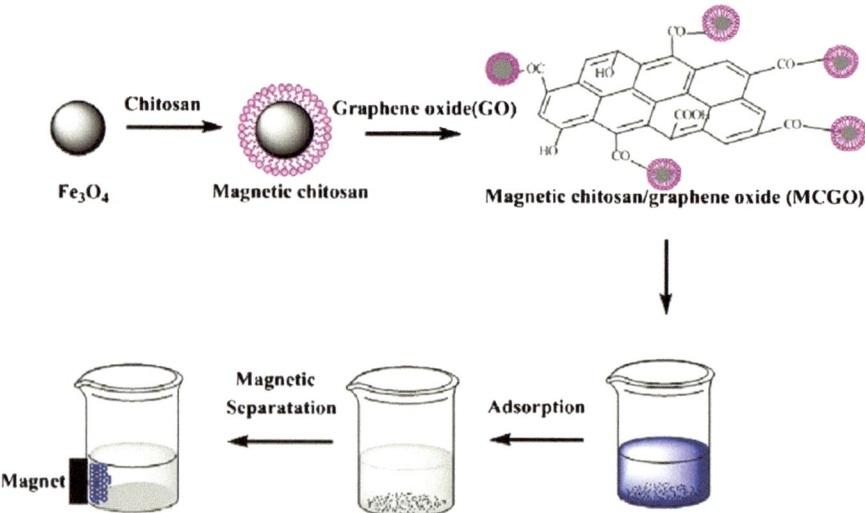

Fig. 6.3 The application of magnetic chitosan/GO for removal of MB with the help of an external magnetic field. Reprinted with permission from Ref. [65]. Copyright (2012) Elsevier Ltd.

mesopore structure, rendering the $Mg(OH)_2$–RGO hybrid an excellent sorbent for MB adsorption [64]. Additionally, the hybrid of magnetic chitosan with GO is also capable to create an external magnetic field for facilitating the MB adsorption [65], as illustrated in Fig. 6.3.

Besides, GO based composites also possess strong adsorption ability for other kinds of organic dyes [66–68]. It was found that the hydroxyl groups play an important role in the acridine orange adsorption on GO, which leads to a extremely high capacity as high as 3,300 mg g^{-1} for the in situ method and 1,400 mg g^{-1} for GO via the Langmuir model [66]. In addition, the 3-D porous GO-polyethylenimine composites exhibited higher adsorption capacity for amaranth (800 mg g^{-1}) far beyond than the orange G and rhodamine B [67]. Also, due to the large surface area and good magnetic responsiveness, the magnetic mesoporous titanium dioxide–graphene oxide (Fe_3O_4-mTiO$_2$-GO) has been utilized as an adsorbent for the removal of Congo Red from simulated wastewater via the aid of an external magnetic field [68]. It reaches a maximum adsorption capacity of 89.95 mg g^{-1}, much higher than other absorbent materials. Most strikingly, the Fe_3O_4-mTiO$_2$-GO composite can be regenerated and reused via simple treatment without any obvious structure and performance degradation.

Furthermore, various GO based composites have been developed for adsorption of co-existing dyes from wastewater [69–73]. Ramesha et al. [69] found that the exfoliated GO (with negatively charged functional groups) and RGO (possessing high surface area) can be used for adsorbing cationic and anionic dyes, respectively. Moreover, the FT-IR and Raman analysis indicate electrostatic and/or van der Waals type interactions between the dyes and the adsorbents. In addition, a

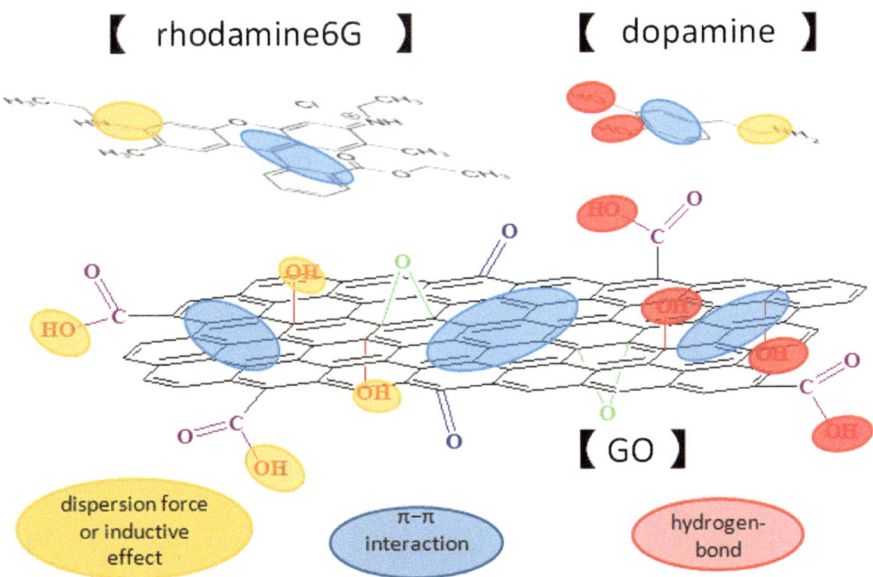

Fig. 6.4 Possible adsorption sites of GO sheets for Rhodamine 6G and dopamine molecules. Reprinted with permission from Ref. [71]. Copyright (2014) American Chemical Society

3-D GO sponge has been employed to simultaneously remove both the MB and methyl violet (MV) dyes [70]. The adsorption process is completed with high efficiency of 99.1 % for MB and 98.8 % for MV in 2 min. The strong π–π stacking and anion-cation interactions between GO sponge and MB/MV ensure superior adsorption capacity as high as 397 and 467 mg g^{-1} for MB and MV dyes, respectively. Moreover, competitive adsorption of Rhodamine 6G (R6G) and dopamine was observed around the oxygen-containing groups of the GO sheets [71], in which the hydrogen bond formation in dopamine-GO (see Fig. 6.4) causes partial desorption of R6G molecules from the GO surface into the solution. Geng et al. [72] reported the application of RGO-Fe$_3$O$_4$ NPs for adsorbing series of dyes such as Rhodamine B, R6G, acid blue 92, orange (II), malachite green and new coccine. They found that this hybrid adsorbent can be easily and efficiently regenerated for reuse by simply annealing in moderate conditions. Moreover, it has almost no adsorption capacity decay for the co-existing dyes. A recent study also demonstrated that the polymethylsiloxane (PDMS) coating on GO can lead to a partially reduced GO/PDMS 3-D architecture with macroporous and hydrophobic properties for remarkably improving the dyes adsorption performance [73].

In practical utility of wastewater treatment, the ideal sorbent needs to be capable of trapping not only the inorganic pollutants but also the organic ones, which motivates the exploration of sorbents materials capable of simultaneously adsorbing the complex pollutants. Accordingly, many GO based materials with various configuration and texture have been developed to meet this require [74–78].

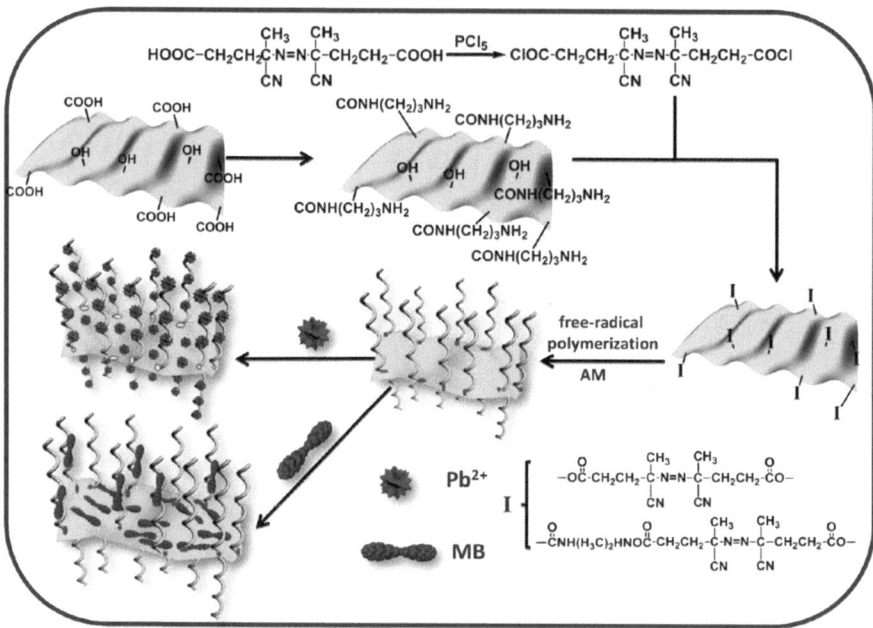

Fig. 6.5 Synthesis of PAM chains on RGO sheets by free radical polymerization and adsorption of Pb(II) and MB. Reprinted with permission from Ref. [76]. Copyright (2013) American Chemical Society

The graphite oxide decorated with thiol groups remains the inherent hydrophilic character and is able to adsorb 6-fold higher concentration of aqueous mercuric ions than the unmodified GO [74]. Moreover, the conversion of regular filtration sand into "core-shell" GO coated sand granules can retain at least 5-fold higher concentration of mercuric ions and RhB than pure sand, which is comparable to some commercially available activated carbon. As shown in Fig. 6.5, the poly(acrylamide) brushes on RGO sheets (RGO/PAM) can convert carboxyl groups into amine groups, leading to the adsorption capacities (via Langmuir isotherm model) as high as 1,000 and 1,530 mg/g for Pb(II) and MB, respectively [76]. In addition, Yang et al. [77] synthesized the iron oxide coated GO and RGO for simultaneously adsorbing Pb(II) and 1-naphthol and 1-naphthylamine. They found that the presence of oxygen-containing groups in GO-iron oxide make the composite a good adsorbent for Pb(II) but not for 1-naphthol and 1-naphthylamine, whereas the RGO-iron oxide material was a good adsorbent for 1-naphthol and 1-naphthylamine but not for Pb(II). Impressively, a very recent study reported the RGO coated by superparamagnetic iron oxide NPs can effectively remove both organic and inorganic pollutants from the contaminated water (for Pb^{2+} and Cd^{2+} within 10 min, whereas for tetrabromobisphenol A within 30 min) [78].

In addition, GO based materials possess well-defined nanometer pores, which allow low frictional water flowing inside and lead to great potential in removing

Fig. 6.6 Sieving through the atomic-scale mesh on the GO membrane. The shown permeation rates are normalized per 1 M feed solution and measured by using 5-mm-thick membranes. Reprinted with permission from Ref. [80]. Copyright (2014) American Association for the Advancement of Science

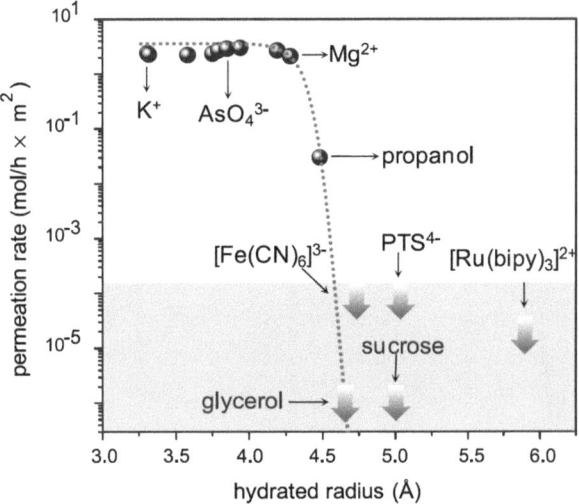

the pollutants from wastewater by selectively ion penetration [79–81]. The free-standing GO membrane fabricated by a simple drop-casting approach was shown to trap the heavy metal ions and organic contaminants [79]. The strong interaction between the RhB molecules and GO membrane and the chemical interactions between the metal ions and the oxygen-containing functional groups account for the selectively penetration phenomenon. Very recently, another study demonstrated the key role of hydrated radii in selectively permeating through the GO membrane [80]. The GO laminates act as molecular sieves and block all solutes with hydrated radii larger than 4.5 Å, whereas smaller ions pass through the membranes with ultrafast ion permeation rate than the case of simple diffusion (see Fig. 6.6).

6.2.2 Conversion of Pollutants

Apart from removal of the pollutants via adsorption and selectively ion penetration, GO based materials also make enormous contribution to the pollutants conversion in water purification technology. For the heavy metal conversion, GO composites show superiors catalytic activity for chemical [44] and photo-reduction [82–84] of toxic Cr(VI) to less toxic Cr(III). Ma et al. [44] found that an ethylenediamine coated RGO (ED-RGO) exhibits a relatively high removal rate of Cr(VI) and can be easily separated from the solution after adsorption. Most strikingly, Cr(VI) can be effectively reduced to low toxic Cr(III) species at low pH, which follows an indirect reduction mechanism with the aid of π electrons on the carbocyclic six-membered ring of ED-RGO. As for the photo-reduction of Cr(VI), TiO_2 NPs decorated RGO (TiO_2-RGO) was used as the photocatalyst to

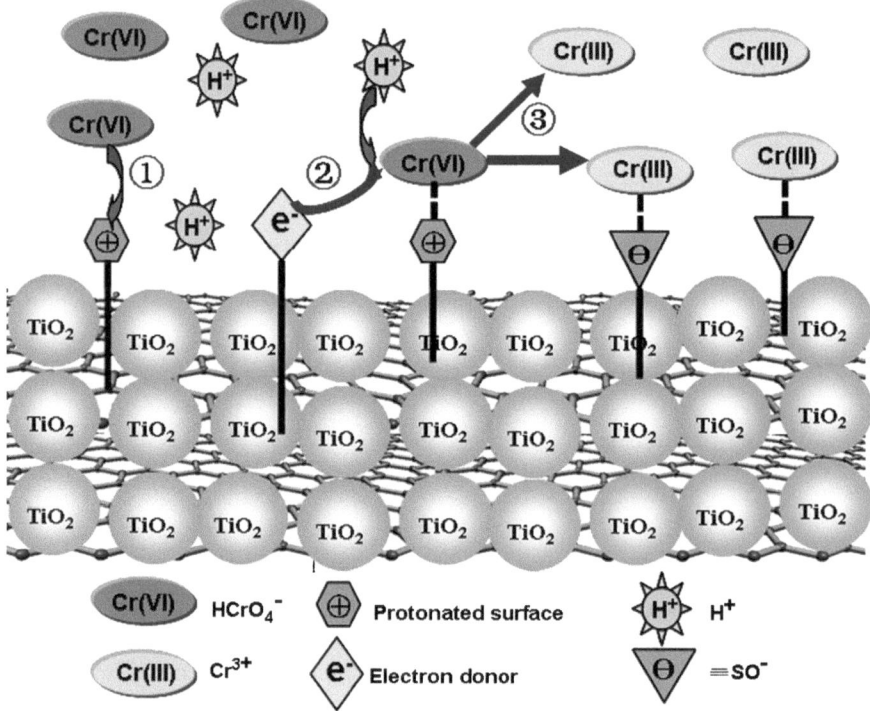

Fig. 6.7 Proposed mechanism of Cr(VI) photocatalytic reduction and removal by TiO₂-RGO. Reprinted with permission from Ref. [82]. Copyright (2013) Elsevier Inc.

remove Cr(VI) from an aqueous solution under visible light irradiation (as shown in Fig. 6.7) [82]. The TiO$_2$-RGO composite exhibits an improved photo-reduction rate of 86.5 % compared to that of 54.2 % for pure TiO$_2$ due to the increased light absorption intensity and wavelength range as well as the reduction in electron-hole pair recombination in TiO$_2$ with the addition of RGO. Similarly, the coupling of RGO with CdS and α-FeOOH nanorod also enhanced the photocatalytic activity of Cr(VI) reduction by the electron-hole pair recombination [83, 84].

GO based composites also function as the efficient catalysts for chemically decomposes the organic pollutants in wastewater [85–88]. The RGO exhibits good catalytic performance for the reduction of nitrobenzene by Na$_2$S in the aqueous solutions, in which the zigzag edges of RGO acts as the catalytic active sites and the basal plane of RGO serves as the conductor for the electron transfer during the catalytic process [85]. Furthermore, changing the pH (5.9–9.1) and the presence of dissolved humic acid (10 mg TOC/L) also intensely affects the catalytic activity of RGO. Oxovanadium(IV) and iron(III) salen complexes coated on the amino-modified GO as a heterogeneous catalyst can significantly enhance the catalytic performance in aerobic oxidation of styrene and retain the cyclic stability compared to its homogeneous analogue [86]. Moreover, it was found that Co$_3$O$_4$

coated GO can effectively catalyze the Orange II decomposition [44, 87] and the Co_3O_4 nanorods coated RGO has great catalytic performance for methylene blue degradation [89].

Besides the chemical catalyst, GO based composites as the photocatalysts have been extensively used for organic pollutants degradation in wastewater. Zhao and coworkers [90] found that RGO can degrade RhB under visible-light irradiation with an extraordinarily slow rate. To improve the photocatalytic performance of RGO for dyes degradation, they fabricated a robust 3-D porous CNT-pillared RGO with excellent visible light photocatalytic activity in the degradation of RhB owing to its unique porous structure and the exceptional electron mobility of graphene [91]. Furthermore, they adopted the Cu modified RGO to enhance the photodegradation of RhB [92].

The metal oxides decorated RGO also exhibit superior photocatalytic performance for the degradation of rhodamine dyes [93–95]. For example, the RhB photodegradation of 98.8 wt% after 80 min for TiO_2–RGO under visible light [93] and the Rh6G photodegradation over 98 % within 10 min for ZnO nanowire/RGO via ultra violet irradiation [94] were observed. The synergistic combination of RGO with other photocatalysts such as Ag_3PO_4 [96] and $BiVO_4$ [97] also lead to significant enhancement of RhB degradation performance. In addition, an Ag-AgCl/rGO nanocomposite (Ag-AgCl encapsulated in GO) has been used as the photocatalyst to degrade RhB through two distinct competing reaction pathways, namely direct electron injection from dye-Ag-AgCl/RGO adsorbents and formation of a surface localized strong oxidant on Ag-AgCl/RGO derived from visible light excitation [98]. Recently, another Ag-AgCl/RGO composite has also been employed to degrade the MB molecules, with the electron-hole pairs of the low energy level recombined in space by metallic Ag as a solid-state electron mediator and the remaining electron-hole pairs of the high energy level for two photochemical reactions that operate in parallel [99].

In addition, RGO composited with other semiconductors like ZnS [100] and metal oxides [101–103] also display enhanced photocatalytic performance of MB degradation compared to that of pure semiconductors. The Ag-AgCl/GO photocatalyst used for metal oxide (MO) degradation has also been documented, in which the GO nanosheets not only control the fabrication of Ag/AgCl nanostructures, but also as catalyst promoter during the photocatalytic performance [104]. The combination of Fe_3O_4-TiO_2 core/shell nanospheres with RGO has also been demonstrated to possess good photocatalytic performance for MO degradation [105]. For the degradation of Cv dye, a nanohybrid of ZnO-GO composite displays high photocatalytic activity (~95 % Cv degraded within 80 min), due to the GO induced high e^--h^+ pair separation under light illumination [106].

In addition to dyes, other organic pollutants can be also degraded by the GO based composites as photocatalyst [107–109]. The ternary catalyst of Ag/RGO–TiO_2 nanotube arrays exhibit nearly 100 % photocatalytic removal efficiency of typical herbicide 2,4-dichlorophenoxyacetic acid (2,4-D) from water under simulated solar light irradiation as well as good cyclic stability over ten times [107]. The UV-cured epoxy films make the GO_x an efficient photocatalyst

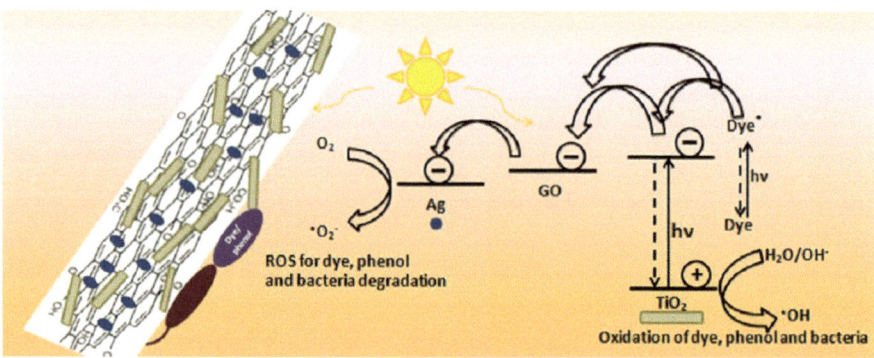

Fig. 6.8 Schematic illustration of the degradation mechanism for dye and phenol. Reprinted with permission from Ref. [112]. Copyright (2013) Elsevier B.V.

for degrading the phenol molecules [108]. Moreover, the CdS nanowires/RGO nanocomposites used for photocatalyst can selectively reduce the aromatic nitro organics in water under visible light irradiation [109].

Most strikingly, some GO based composites can simultaneously degrade various organic pollutants [110–114], which is favorable in the practical water purification. For instance, the GO enwrapped Ag_3PO_4 composite exhibits enhanced photocatalytic activities and cyclic stability for the degradation of organic dye (AO7) and phenol under visible-light irradiation compared to the pure Ag_3PO_4 NPs [110]. Later, another GO/Ag_3PO_4 composite shows great potential in photodegradation of RhB and bisphenol A [111]. Liu et al. [112] further found that the ternary GO-TiO_2-Ag nanocomposites remarkably enhance the photocatalytic activities in degrading AO7 and phenol under solar irradiation compared with GO/TiO_2 and GO/Ag. The optimal Ag content in the GO/TiO_2/Ag nanocomposites varies for dye degradation and for phenol/bacterial degradation with different mechanisms, as illustrated in Fig. 6.8. Additionally, the combination of RGO with β-$SnWO_4$ improves the degradation efficiency of MO and RhB from 55 and 60 % up to 90 and 91 %, respectively, compared to the β-$SnWO_4$ alone [113].

References

1. Zhao, Y.X., Ding, H.L., Zhong, Q.: Appl. Surf. Sci. **258**, 4301–4307 (2012)
2. Kim, H.W., Yoon, H.W., Yoon, S.M., Yoo, B.M., Ahn, B.K., Cho, Y.H., Shin, H.J., Yang, H., Paik, U., Kwon, S., Choi, J.Y., Park, H.B.: Science **342**, 91–95 (2013)
3. Eigler, S., Dotzer, C., Hirsch, A., Enzelberger, M., Müller, P.: Chem. Mater. **24**, 1276–1282 (2012)
4. Kim, D., Kim, D.W., Lim, H.K., Jeon, J., Kim, H.J., Jung, H.T., Lee, H.: J. Phys. Chem. C **118**, 11142–11148 (2014)
5. Yumura, T., Yamasaki, A.: Phys. Chem. Chem. Phys. **16**, 9656–9666 (2014)
6. Garcia-Gallastegui, A., Iruretagoyena, D., Gouvea, V., Mokhtar, M., Asiri, A.M., Basahel, S.N., Al-Thabaiti, S.A., Alyoubi, A.O., Chadwick, D., Shaffer, M.S.P.: Chem. Mater. **24**, 4531–4539 (2012)

7. Iruretagoyena, D., Shaffer, M.S.P., Chadwick, D.: Adsorption **20**, 321–330 (2014)
8. Kemp, K.C., Chandra, V., Saleh, M., Kim, K.S.: Nanotechnology **24**, 235703–235710 (2013)
9. Saleh, M., Chandra, V., Kemp, K.C., Kim, K.S.: Nanotechnology **24**, 255702–255709 (2013)
10. Alhwaige, A.A., Agag, T., Ishida, H., Qutubuddin, S.: RSC Adv. **3**, 16011–16020 (2013)
11. Liu, S., Sun, L.X., Xu, F., Zhang, J., Jiao, C.L., Li, F., Li, Z.B., Wang, S., Wang, Z.Q., Jiang, X., Zhou, H.Y., Yang, L.N., Schick, C.: Energy Environ. Sci. **6**, 818–823 (2013)
12. Garcia-Gallastegui, A., Iruretagoyena, D., Mokhtar, M., Asiri, A.M., Basahel, S.N., Al-Thabaiti, S.A., Alyoubi, A.O., Chadwick, D., Shaffer, M.S.P.: J. Mater. Chem. **22**, 13932–13940 (2012)
13. Seredych, M., Bandosz, T.J.: J. Phys. Chem. C **111**, 15596–15604 (2007)
14. Petit, C., Seredych, M., Bandosz, T.J.: J. Mater. Chem. **19**, 9176–9185 (2009)
15. Seredych, M., Bandosz, T.J.: Langmuir **26**, 5491–5498 (2010)
16. Seredych, M., Rossin, J.A., Bandosz, T.J.: Carbon **49**, 4392–4402 (2011)
17. Petit, C., Bandosz, T.J.: J. Phys. Chem. C **113**, 3800–3809 (2009)
18. Seredych, M., Bandosz, T.: J. Colloids Surf., A **353**, 30–36 (2010)
19. Seredych, M., Bandosz, T.: J. Mesoporous Mater. **150**, 55–63 (2012)
20. Tang, S.B., Cao, Z.X.: J. Phys. Chem. C **116**, 8778–8791 (2012)
21. Mattson, E.C., Pande, K., Unger, M., Cui, S., Lu, G., Gajdardziska-Josifovska, M., Weinert, M., Chen, J.H., Hirschmugl, C.J.: J. Phys. Chem. C **117**, 10698–10707 (2013)
22. Matsuo, Y., Nishino, Y., Fukutsuka, T., Sugie, Y.: Carbon **46**, 1159–1174 (2008)
23. He, Y.Q., Zhang, N.N., Wu, F., Xu, F.Q., Liu, Y., Gao, J.P.: Mater. Res. Bull. **48**, 3553–3558 (2013)
24. Huang, Z.H., Liu, G.Q., Kang, F.Y.: ACS Appl. Mater. Interfaces **4**, 4942–4947 (2012)
25. Huang, L.L., Seredych, M., Bandosz, T.J., van Duin, A.C.T., Lu, X.H., Gubbins, K.E.: J. Chem. Phys. **139**, 194707–194715 (2013)
26. Seredych, M., Bandosz, T.J.: J. Phys. Chem. C **114**, 14552–14560 (2010)
27. Wang, L., Zhao, J.J., Wang, L.L., Yan, T.Y., Sun, Y.Y., Zhang, S.B.: Phys. Chem. Chem. Phys. **13**, 21126–21131 (2011)
28. Tang, S.B., Cao, Z.X.: J. Chem. Phys. **134**, 23–044710 (2011)
29. Chen, C., Xu, K., Ji, X., Miao, L., Jiang, J.J.: Phys. Chem. Chem. Phys. **16**, 11031–11036 (2014)
30. He, F., Niu, N., Qu, F.Y., Wei, S.Q., Chen, Y.J., Gai, S.L., Gao, P., Wang, Y., Yang, P.P.: Nanoscale **5**, 8507–8516 (2013)
31. Das, T., Deo, G.: Catal. Today **198**, 116–124 (2012)
32. Lan, D.H., Yang, F.M., Luo, S.L., Au, C.T., Yin, S.F.: Carbon **73**, 351–360 (2014)
33. An, X.Q., Li, K.F., Tang, J.W.: ChemSusChem **7**, 1086–1093 (2014)
34. Cheng, J., Zhang, M., Wu, G., Wang, X., Zhou, J.H., Cen, K.F.: Environ. Sci. Technol. **48**, 7076–7084 (2014)
35. Hsu, H.C., Shown, I., Wei, H.U., Chang, Y.C., Du, H.Y., Lin, Y.G., Tseng, C.A., Wang, C.H., Chen, L.C., Lind, Y.C., Chen, K.H.: Nanoscale **5**, 262–268 (2013)
36. Kumar, P., Kumar, A., Sreedhar, B., Sain, B., Ray, S.S., Jain, S.L.: Chem. Eur. J. **20**, 6154–6161 (2014)
37. Tan, L.L., Ong, W.J., Chai, S.P., Mohamed, A.R.: Nanoscale Res. Lett. **8**, 465–473 (2013)
38. Yu, J.G., Jin, J., Cheng, B., Jaroniec, M.: J. Mater. Chem. A **2**, 3407–3416 (2014)
39. Long, Y., Zhang, C.C., Wang, X.X., Gao, J.P., Wang, W., Liu, Y.: J. Mater. Chem. **21**, 13934–13941 (2011)
40. Humeres, E., Debacher, N.A., Smaniotto, A., Castro, K.M.D., Benetoli, L.O.B., Souza, E.P.D., Moreira, R.D.F.P.M., Lopes, C.N., Schreiner, W.H., Canle, M., Santaballa, J.A.: Langmuir **30**, 4301–4309 (2014)
41. Nie, R.F., Shi, J.J., Xia, S.X., Shen, L., Chen, P., Hou, Z.Y., Xiao, F.S.: J. Mater. Chem. **22**, 18115–18118 (2012)
42. Andryushina, N.S., Stroyuk, O.L.: Appl. Catal. B**148–149**, 543–549 (2014)
43. Jia, T.T., Lu, C.H., Zhang, Y.F., Chen, W.K.: J. Nanopart. Res. **16**, 2206–2216 (2014)

44. Ma, H.L., Zhang, Y., Hu, Q.H., Yan, D., Yu, Z.Z., Zhai, M.: J. Mater. Chem. **22**, 5914–5916 (2012)
45. Dinda, D., Gupta, A., Saha, S.K.: J. Mater. Chem. A **1**, 11221–11228 (2013)
46. Li, S.K., Lu, X.F., Xue, Y.P., Lei, J.Y., Zheng, T., Wang, C.: PLoS ONE **7**, 43328–43334 (2012)
47. Zhang, S.W., Zeng, M.Y., Xu, W.Q., Li, J.X., Li, J., Xu, J.Z., Wang, X.K.: Dalton Trans. **42**, 7854–7858 (2013)
48. Ansari, R., Fahim, N.K.: React. Funct. Polym. **67**, 367–374 (2007)
49. Jabeen, H., Chandra, V., Jung, S., Lee, J.W., Kim, K.S., Kim, S.B.: Nanoscale **3**, 3583–3585 (2011)
50. Yao, T.J., Cui, T.Y., Wu, J., Chen, Q.Z., Lu, S.W., Sun, K.N.: Polym. Chem. **2**, 2893–2899 (2011)
51. Liu, M.C., Chen, C.L., Hu, J., Wu, X.L., Wang, X.K.: J. Phys. Chem. C **115**, 25234–25240 (2011)
52. Yang, S.T., Chang, Y.L., Wang, H.F., Liu, G.B., Chen, S., Wang, Y.W., Liu, Y.F., Cao, A.N.: J. Colloid Interface Sci. **351**, 122–127 (2010)
53. Mi, X., Huang, G.B., Xie, W.S., Wang, W., Liu, Y., Gao, J.P.: Carbon **50**, 4856–4864 (2012)
54. Zhao, G.X., Ren, X.M., Gao, X., Tan, X.L., Li, J.X., Chen, C.L., Huang, Y.Y., Wang, X.K.: Dalton Trans. **40**, 10945–10952 (2011)
55. Madadrang, C.J., Kim, H.Y., Gao, G.H., Wang, N., Zhu, J., Feng, H., Gorring, M., Kasner, M.L., Hou, S.F.: ACS Appl. Mater. Interfaces **4**, 1186–1193 (2012)
56. Chandra, V., Kim, K.S.: Chem. Commun. **47**, 3942–3944 (2011)
57. Zhang, N.N., Qiu, H.X., Si, Y.M., Wang, W., Gao, J.P.: Carbon **49**, 827–837 (2011)
58. Lee, Y.C., Yang, J.W.: J. Ind. Eng. Chem. **18**, 1178–1185 (2012)
59. Liu, L., Liu, S.X., Zhang, Q.P., Li, C., Bao, C.L., Liu, X.T., Xiao, P.F.: J. Chem. Eng. Data **58**, 209–216 (2013)
60. Zhou, L., Gao, C., Xu, W.J.: ACS Appl. Mater. Interfaces **2**, 1483–1491 (2010)
61. Yang, S.T., Chen, S., Chang, Y.L., Cao, A.N., Liu, Y.F., Wanga, H.F.: J. Colloid Interface Sci. **359**, 24–29 (2011)
62. Chen, L.J., Yang, J.J., Zeng, X.P., Zhang, L.Z., Yuan, W.X.: Mater. Express **3**, 281–290 (2013)
63. He, G.Y., Zhang, J.G., Zhang, Y., Chen, H.Q., Wang, X.: J. Dispersion Sci. Technol. **34**, 1223–1229 (2013)
64. Li, B.J., Cao, H.Q., Yin, G.: J. Mater. Chem. **21**, 13765–13768 (2011)
65. Fan, L.L., Luo, C.N., Sun, M., Li, X.J., Lu, F.G., Qiu, H.M.: Bioresour. Technol. **114**, 703–706 (2012)
66. Sun, L., Yu, H.W., Fugetsu, B.: J. Hazard. Mater. **203–204**, 101–110 (2012)
67. Sui, Z.Y., Cui, Y., Zhu, J.H., Han, B.H.: ACS Appl. Mater. Interfaces **5**, 9172–9179 (2013)
68. Li, L.L., Li, X.J., Duan, H.M., Wang, X.X., Luo, C.N.: Dalton Trans. **43**, 8431–8438 (2014)
69. Ramesha, G.K., Kumara, A.V., Muralidhara, H.B., Sampath, S.: J. Colloid Interface Sci. **361**, 270–277 (2011)
70. Liu, F., Chung, S., Oh, G., Seo, T.S.: ACS Appl. Mater. Interfaces **4**, 922–927 (2012)
71. Ren, H., Kulkarni, D.D., Kodiyath, R., Xu, W.N., Choi, I., Tsukruk, V.V.: ACS Appl. Mater. Interfaces **6**, 2459–2470 (2014)
72. Geng, Z.G., Lin, Y., Yu, X.X., Shen, Q.H., Ma, L., Li, Z.Y., Pan, N., Wang, X.P.: J. Mater. Chem. **22**, 3527–3535 (2012)
73. Park, S., Kang, S.O., Jung, E., Sungyoul Park, S., Park, H.S.: RSC Adv. **4**, 899–902 (2014)
74. Gao, W., Majumder, M., Alemany, L.B., Narayanan, T.N., Ibarra, M.A., Pradhan, B.K., Pulickel, M., Ajayan, P.M.: ACS Appl. Mater. Interfaces **3**, 1821–1826 (2011)
75. Yang, Z., Yan, H., Yang, H., Li, H.B., Li, A.M., Cheng, R.: Water Res. **47**, 3037–3046 (2013)
76. Yang, Y.F., Xie, Y.L., Pang, L.C., Li, M., Song, X.H., Wen, J.G., Zhao, H.Y.: Langmuir **29**, 10727–10736 (2013)
77. Yang, X., Chen, C.L., Li, J.X., Zhao, G.X., Ren, X.M., Wang, X.K.: RSC Adv. **2**, 8821–8826 (2012)
78. Thakur, S., Karak, N.: Mater. Chem. Phys. **144**, 425–432 (2014)
79. Sun, P.Z., Zhu, M., Wang, K.L., Zhong, M.L., Wei, J.Q., Wu, D.H., Xu, Z.P., Zhu, H.W.: ACS Nano **7**, 428–437 (2013)
80. Joshi, R.K., Carbone, P., Wang, F.C., Kravets, V.G., Su, Y., Grigorieva, I.V., Wu, H.A., Geim, A.K., Nair, R.R.: Science **343**, 752–754 (2014)

81. Xu, C., Cui, A.J., Xu, Y.L., Fu, X.Z.: Carbon **62**, 465–471 (2013)
82. Zhao, Y., Zhao, D.L., Chen, C.L., Wang, X.K.: J. Colloid Interface Sci. **405**, 211–217 (2013)
83. Pawar, R.C., Lee, C.C.: Mater. Chem. Phys. **141**, 686–693 (2013)
84. Padhi, D.K., Parida, K.: J. Mater. Chem. A **2**, 10300–10312 (2014)
85. Fu, H.Y., Zhu, D.Q.: Environ. Sci. Technol. **47**, 4204–4210 (2013)
86. Li, Z.F., Wu, S.J., Ding, H., Lu, H.M., Liu, J.Y., Huo, Q.S., Guan, J.Q., Kan, Q.B.: New J. Chem. **37**, 4220–4229 (2013)
87. Shi, P.H., Su, R.J., Zhu, S.B., Zhu, M.C., Li, D.X., Xu, S.H.: J. Hazard. Mater. **229–230**, 331–339 (2012)
88. Shi, P.H., Dai, X.F., Zheng, H.G., Li, D.X., Yao, W.F., Hu, C.Y.: Chem. Eng. J. **240**, 264–270 (2014)
89. Zhang, Z., Hao, J.H., Yang, W.S., Lu, B.P., Ke, X., Zhang, B.L., Tang, J.L.: ACS Appl. Mater. Interfaces **5**, 3809–3815 (2013)
90. Xiong, Z.G., Zhang, L.L., Ma, J., Zhao, X.S.: Chem. Commun. **46**, 6099–6101 (2010)
91. Zhang, L.L., Xiong, Z.G., Zhao, X.S.: ACS Nano **4**, 7030–7036 (2010)
92. Xiong, Z.G., Zhang, L.L., Zhao, X.S.: Chem. Eur. J. **17**, 2428–2434 (2011)
93. Shah, M.S.A.S., Park, R., Zhang, K., Park, J.H., Yoo, P.J.: ACS Appl. Mater. Interfaces **4**, 3893–3901 (2012)
94. Zhang, Z., Zhang, J., Su, Y.J., Xu, M.H., Zhang, Z., Zhang, Y.F.: Physica E **56**, 251–255 (2014)
95. Liu, H., Liu, T.T., Dong, X.N., Lv, Y.W., Zhu, Z.F.: Mater. Lett. **126**, 36–38 (2014)
96. Chai, B., Li, J., Xu, Q.: Ind. Eng. Chem. Res. **53**, 8744–8752 (2014)
97. Dong, S.Y., Cui, Y.R., Wang, Y.F., Li, Y.K., Hu, L.M., Sun, J.Y., Sun, J.H.: Chem. Eng. J. **249**, 102–110 (2014)
98. Miller, C.J., Yu, H.J., Waite, T.D.: Colloids Surf., A **435**, 147–153 (2013)
99. Min, Y.L., He, G.Q., Xu, Q.J., Chen, Y.C.: J. Mater. Chem. A **2**, 1294–1301 (2014)
100. Sookhakiana, M., Amina, Y.M., Basirun, W.J.: Appl. Surf. Sci. **283**, 668–677 (2013)
101. Nguyen, P.T., Salim, C., Kurniawan, W., Hinode, H.: Catal. Today **230**, 166–173 (2014)
102. Ariffin, S.N., Lim, H.N., Jumeri, F.A., Zobir, M., Abdullah, A.H., Ahmad, M., Ibrahim, N.A., Huang, N.M., Teo, P.S., Muthoosamy, K., Harrison, I.: Ceram. Int. **40**, 6927–6936 (2014)
103. Huang, K., Li, Y.H., Lin, S., Liang, C., Xu, X., Zhou, Y.F., Fan, D.Y., Yang, H.J., Lang, P.L., Zhang, R., Wang, Y.G., Lei, M.: Mater. Lett. **124**, 223–226 (2014)
104. Zhu, M.S., Chen, P.L., Liu, M.H.: Langmuir **29**, 9259–9268 (2013)
105. Ma, P.C., Jiang, W., Wang, F.H., Li, F.S., Shen, P., Chen, M.D., Wang, Y., Liu, J., Li, P.Y.: J. Alloy. Compd. **578**, 501–506 (2013)
106. Ameen, S., Akhtar, M.S., Seo, H.K., Shin, H.S.: Mater. Lett. **100**, 261–265 (2013)
107. Tang, Y.H., Luo, S.L., Teng, Y.R., Liu, C.B., Xu, X.L., Zhang, X.L.: J. Hazard. Mater. **241–242**, 323–330 (2012)
108. Sangermano, M., Calza, P., Lopez-Manchado, M.A.: J. Mater. Sci. **48**, 5204–5208 (2013)
109. Liu, S.Q., Chen, Z., Zhang, N., Tang, Z.R., Xu, Y.J.: J. Phys. Chem. C **117**, 8251–8261 (2013)
110. Liu, L., Liu, J.C., Sun, D.D.: Catal. Sci. Technol. **2**, 2525–2532 (2012)
111. Wang, C., Zhu, J.X., Wu, X.Y., Xu, H., Song, Y.H., Yan, J., Song, Y.X., Ji, H.Y., Wang, K., Li, H.M.: Ceram. Int. **40**, 8061–8070 (2014)
112. Liu, L., Bai, H.W., Liu, J.C., Sun, D.D.: J. Hazard. Mater. **261**, 214–223 (2013)
113. Liu, H., Chen, Z., Jin, Z.T., Su, Y., Wang, Y.: Dalton Trans. **43**, 7491–7498 (2014)
114. Thangavel, S., Venugopal, G., Kim, S.J.: Mater. Chem. Phys. **145**, 108–115 (2014)

Chapter 7
Application of GO in Biotechnology

Abstract The functional groups on GO, such as hydroxyl, epoxide and carboxyl, make GO hydrophilic and dissolvable due to the high affinity of these groups to water molecules. The functional groups also allow GO to noncovalently interact with biomolecules via electrostatic interaction, π–π stacking, and hydrogen bonding. Therefore, GO with hydrophilic/active surface and high surface area is a promising material in biotechnology. Moreover, the excellent optical and electromechanical properties of GO extend its applications in biotechnology, especially as biosensors, which can be used to detect enzyme, DNA and other biomolecules with high sensitivity and selectivity.

7.1 Biofunctionalization

Recently, applications of graphene and its derivatives (chemically modified graphene, GO, etc.) in biotechnology have attracted great attentions. Various biomolecules have been reported to be integrated with graphene-based nanomaterials [1], such as proteins, nucleic acids, cells, bacteria, and so on, as illustrated in Fig. 7.1.

7.1.1 Interaction with Proteins

Owing to its biocompatibility, GO can be biofunctionalized with many kinds of proteins through physical adsorption or chemical bonding. For instance, enzyme, such as horseradish peroxidase (HRP) and lysozyme, can be easily immobilized on GO sheet through electrostatic interaction without any cross-linking reagents or additional surface modification since GO is enriched with oxygen-containing groups [2]. Moreover, the flat surface of GO makes it possible to observe the native immobilized enzyme in situ with AFM, as shown in Fig. 7.2. Note that the

© The Author(s) 2015
J. Zhao et al., *Graphene Oxide: Physics and Applications*,
SpringerBriefs in Physics, DOI 10.1007/978-3-662-44829-8_7

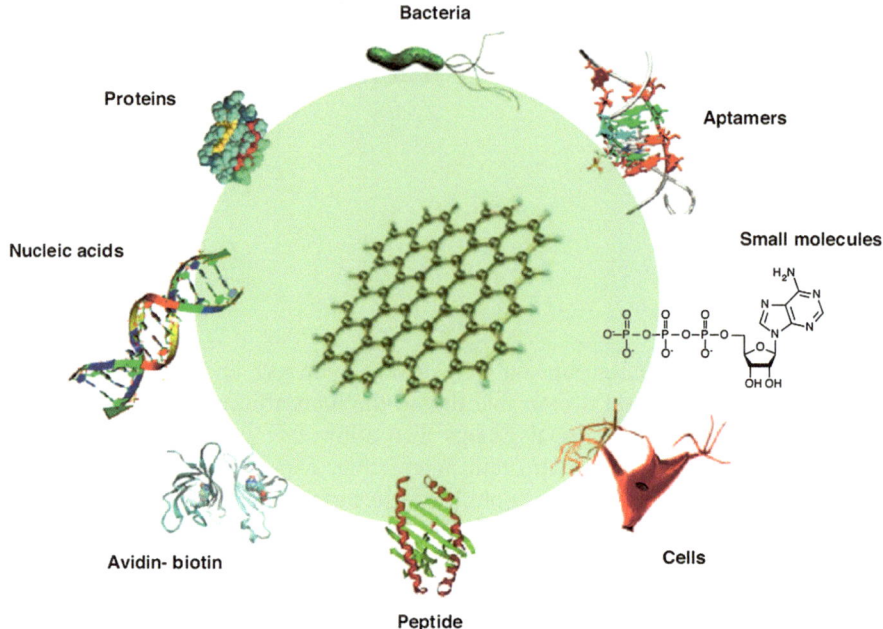

Fig. 7.1 Biomolecules reported to be functionalized with graphene and its derivatives through physical adsorption or chemical conjugation. Reprinted with permission from Ref. [1]. Copyright (2011) Elsevier Ltd.

immobilized enzyme is mainly determined by the interaction of enzyme molecules with the functional groups of GO [2]. Using sodium dodecyl sulfate–polyacrylamide gel electrophoresis (SDS–PAGE), Kotchey et al. [3] also confirmed the ability of HRP to bind with the RGO with an overall binding energy of −26.7 kcal/mol. Meanwhile, the HRP is preferentially bound to the basal plane rather than the edge for both GO and RGO.

Another important enzyme that can be biofunctionalized with GO is glucose oxidase (GOD), and the resulting complex can serve as electrochemical glucose biosensors, as covered in a recent review [4]. The RGO can be modified with GOD by covalent bonding via a polymer generated by electrografting N-succinimidyl acrylate (NSA) [5]. The direct electron transfer between the RGO-based electrode and GOD molecules is realized, which can be utilized to detect glucose. Also, GO can interact with GOD through covalent attachment between carboxyl acid groups of GO sheets and amines of GOD [6]. Similarly, weak electron transferring nature of GOD was observed, which can further decrease the electrochemical performance of the GO electrode. The covalently linked GOD-GO enzyme electrode shows broad linearity, good sensitivity, excellent reproducibility and storage stability, suggesting GO to be a highly efficient biosensor electrode.

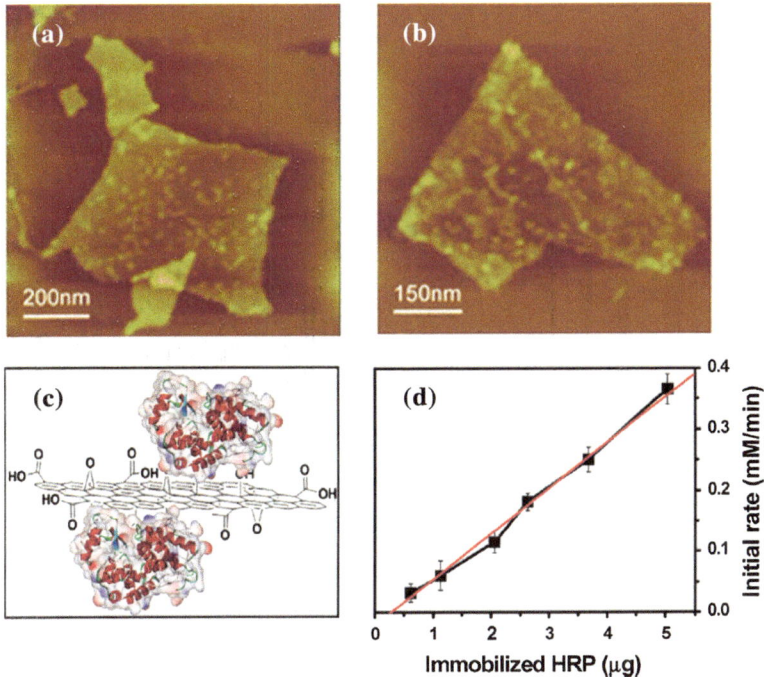

Fig. 7.2 AFM images of the GO-bound HRP with lower (**a**) and higher (**b**) enzyme loadings acquired in a liquid cell. **c** Schematic model of the GO-bound HRP. **d** Initial reaction rates of GO-bound HRP versus HRP concentration. Reprinted with permission from Ref. [2]. Copyright (2010) American Chemical Society

7.1.2 Interaction with DNA

In addition to proteins, GO can be also biofunctionalized with DNA. Through the strong π–π stacking force, single-stranded DNA (ssDNA) can be decorated on the basal plane of GO [7–9]. Due to the high planar surface of GO, adsorption of ssDNA occurs in only few minutes since more than 95 % fluorescence intensities are quenched within 3 min, which is faster than using single walled carbon nanotubes [10]. Benefiting from the π–π reaction between nucleotide bases and GO, the ssDNA–GO complex demonstrates superior sensitivity and rapid response.

Liu et al. [7] also investigated decoration of the ssDNA on GO. Generally, there are two kinds of driving forces for the adsorption of ssDNA. One is the strong π–π stacking interaction between the ssDNA and the graphene-like local conjugated domains on the GO surface. Another is the hydrogen bonding between ssDNA and some oxygen-containing groups on GO. Moreover, after adsorption of the ssDNA on the GO surface, the resulting ssDNA–GO bioconjugate bears multiple thiol groups tagged on DNA strands and can be used as a 2-D bionanointerface for assembling gold nanoparticles.

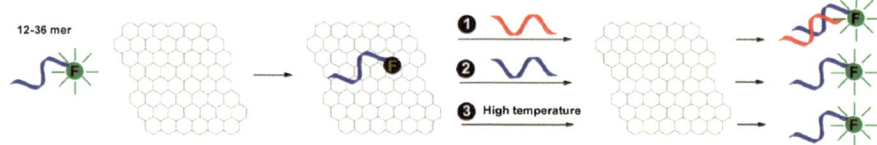

Fig. 7.3 Schematic showing FAM-labeled DNA desorption on GO. Fluorescence is quenched upon adsorption. Desorption can be achieved by adding the c-DNA (*reaction 1*), DNA exchange with the same DNA (*reaction 2*), or increasing temperature (*reaction 3*). Reprinted with permission from Ref. [9]. Copyright (2011) American Chemical Society

Similarly, Xu et al. [8] reported a strategy for 3-D self-assembly of GO sheets and ssDNA via strong noncovalent interactions to form multifunctional hydrogels with high mechanical strength, excellent environmental stability, high dye-adsorption capacity, and self-healing function. Then, Wu et al. [9] systematically investigated the interaction between GO and DNA. They found that the adsorption and desorption of DNA on GO are affected by several factors, such as cations, pH, organic solvent and temperature. Generally, shorter DNA binds to the surface with faster kinetics and higher adsorption efficiency. However, by adding the c-DNA to form double-stranded (dsDNA) and increasing temperature, desorption can occur, as shown in Fig. 7.3.

On the other hand, GO scarcely interact with the rigid structure of dsDNA probably due to the efficient shielding of nucleobases within the negatively charged phosphate backbone of dsDNA [11]. According to the fluorescence quenching of DNA–GO complex, it was found that the formation and release process of the dsDNA from GO is relatively slow [12]. He et al. [13] also confirmed that interactions between the dsDNA and GO are rather weak from analysis of the fluorescence spectra of the dsDNA–GO complex because the FAM (carboxy fluorescein) fluorescence is largely remained in the presence of GO. Meanwhile, fluorescence intensity of the dsDNA can be about 50 times larger than that of the ssDNA in the presence of T1 (5′-CAGACAAACTCCAACGA-3′) at 5-fold excess (50 nM). In addition, Dong et al. [14] suggested that the formation of dsDNA–quantum dots (QDs) will reduce the surface charge of the DNA molecules and the exposure of the base. The latter effect can weaken the π–π stacking and hydrogen bonding interactions, further leading to a decrease in the adsorption rate of dsDNA–QDs on GO.

7.1.3 Interaction with Other Biomolecules

Other biomolecules, including hydrogen peroxide (H_2O_2), β-nicotinamide adenine dinucleotide (NADH), dopamine (DA), peptides and cellulose, have also been biofunctionalized with GO [15, 16]. H_2O_2 is a general enzymatic product of oxidases and NADH is a cofactor of many dehydrogenases, both of which are important in biological processes and biosensor development [15]. It was found that the

H$_2$O$_2$- and NADH-functionalized CRGO sheets can be used as bioelectrodes with greatly enhanced electrochemical reactivity [16]. Then, Tang et al. [17] attributed the enhanced electrochemical behavior of the NADH–GO modified electrode to the high density of edge-plane-like defective sites on the RGO, which provide many active sites for electron transfer to the biological species.

DA, also known as a hormone and neurotransmitter, is a unique molecule mimicking the adhesive proteins. Xu et al. [18] suggested DA can also be employed to immobilize thiol- and amino-terminated poly ethylene glycol (PEG) on the surface of RGO in a "grafting-to" process. Meanwhile, DA also allows the reduction of GO without usage of hazardous chemicals or reducing agents. Peptides are short chains of amino acid monomers linked by peptide (amide) bonds, which allow the creation of peptide antibodies in animals without the need to purify the protein of interest [19]. Due to the electrostatic attraction between RGO sheets and protonated peptides, peptide-GO hybrid can be assembled into core-hell nanowires through a simple solution mixing [20].

Cellulose is an important structural component of the primary cell wall of green plants [21]. Moreover, cellulose derivatives are biologically compatible and biodegradable natural polymeric dispersants, which have been employed to fabricate high-concentration and stable aqueous suspensions of graphene nanosheets through chemical reduction of exfoliated GO [22]. It was found that the resulting suspensions of the cellulose-GO nanosheets are very stable in water even at high concentrations (0.6–2 mg/ml) [22].

7.2 Bioapplications

7.2.1 Optical Biosensors

As mentioned in Chaps. 3 and 4, GO has fluorescence in the near-infrared to ultraviolet regions. One important application of GO is to act as donor or acceptor of Förster (or fluorescence) resonance energy transfer (FRET). FRET is a phenomenon in which photo excitation energy is transferred from a donor fluorophore to an acceptor molecule [11].

First, GO can serve as a donor of FRET. Taking GO sheets as donors and gold nanoparticles (AuNPs) as acceptors of FRET, the complex can be used to detect a rotavirus [23] and DNA [24] with high sensitivity and selectivity by GO photoluminescence quenching. To detect the rotavirus, AuNP-linked antibodies bridged with 100-mer ssDNA molecules, i.e. Ab-DNA-AuNP complexes, have been fabricated, as shown in Fig. 7.4a. The ssDNA molecule is used as a mediator to facilely control the distance between antibodies and AuNPs so that the AuNPs can be placed close to the GO surface. When the Ab-DNA-AuNP complexes are selectively bound to the target cells that are attached to the GO arrays, a reduction in the fluorescence emission of GO by quenching is detected, which enables the identification of pathogenic target cells, as displayed in Fig. 7.4b [23].

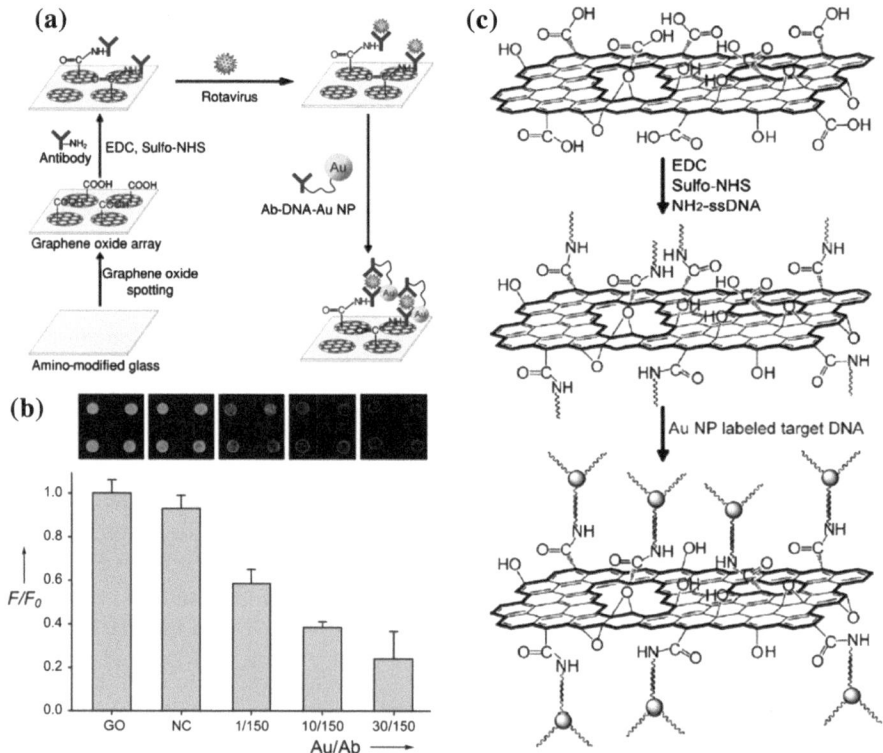

Fig. 7.4 Illustration of a GO-based immuno-biosensor (**a**) and its photoluminescence quenching effect (**b**). Reprinted with permission from Ref. [23]. Copyright (2010) Wiley-VCH Verlag GmbH & Co. KGaA, Weinheim. **c** Schematic diagram to show the GO-based DNA biosensor. Reprinted with permission from Ref. [24]. Copyright (2010) Elsevier B.V.

Similarly, GO in an array format can be used as a novel DNA biosensor to recognize the specific DNA–DNA hybridization interactions, as illustrated in Fig. 7.4c [24]. By using carbodiimide chemistry, the DNA probe is linked to the surface of GO. Then the GO-linked DNA is further hybridized with an AuNP-labeled complementary DNA strand. In the AuNP-dsDNA-GO complex, the closely located GO and AuNPs act as energy donor and acceptor, respectively, which leads to quenching of GO fluorescence through the FRET phenomenon. It is noteworthy that the fluorescence emission intensity of the GO array is drastically reduced due to the quenching of fluorescence energy transfer between AuNPs and the GO sheets.

Besides, through covalent binding, Song et al. was able to construct a GO-based FRET sensor for sensitive, rapid, and accurate detection of matrix metalloproteinases (MMPs) in complex serum samples. In the GO-based FRET sensor, fluorescein isothiocyanate-labeled peptide (Pep-FITC) was assembled onto the GO surface [25]. It was reported that the GO-Pep-FITC FRET sensor achieves rapid MMPs detection within 3 h even in complex biological samples with satisfactory accuracy and small relative standard deviation ($\leq$7.03 %).

On the other hand, GO is able serve as an acceptor of FRET. Lu et al. [12] pointed out that GO can bind dye-labeled ssDNA and completely quench the fluorescence of the dye through strong noncovalent π–π stacking interactions. However, the presence of a target will change the conformation of dye-labeled DNA, which further disturbs the interaction between the dye-labeled DNA and GO. Interruption of the interaction will release the dye-labeled DNA from the GO, resulting in restoration of dye fluorescence. Consequently, this strategy results in a fluorescence-enhanced detection that is sensitive and selective to the target molecule. Due to the highly planar surface and universal quenching capabilities of GO, this mechanism can be extended to a multiplexed (multicolor) DNA detection [13] or a multiplexed detection of different targets (DNA, proteins, metal ions, etc.) [10]. For example, QD nanocrystal quenching by GO has been exploited for DNA detection by Dong et al. [14]. Using molecular beacon, the QDs are quenched onto the GO surface by π–π stacking interactions. During hybridization process, the quenched QDs can be detached from the GO, recovering the fluorescence of QDs.

For protein detection, Liu et al. [26] have designed a promising self-assembled homogeneous immunoassay to detect trace biomarker protein with distance-independent quenching efficiency, which is based on modulating the interaction between GO sheets and inorganic luminescent QDs. When adding the target AFP to the mixture of reporter antibody labeled QDs and capture antibody modified GO sheets in solution, the emission of QDs is quenched by performing sandwich immunocomplexes in a single step. Moreover, the radiative quenching efficiency is distance-independent in a wide dynamic range due to the effect of the graphene-based material, which significantly breaks the distance limit in traditional FRET-based biosensors.

Due to the high quenching efficiency of GO, Wang et al. [27] demonstrated that GO is a better FRET acceptor compared to SWNTs for cyclin A_2 detection according to Stern–Volmer relationship measurements. Using GO, the reported detection limit is 0.5 nM, ten times better than that with SWNTs. Table 7.1 summarizes GO as an acceptor to biosense various biomolecules, where E_{xc} is the excitation wavelength; E_m is the maximum emission wavelength; MQE is the maximum quenching efficiency within the assay; LOD is the limit of detection; FAM is the 6-carboxyfluorescein; PNBP is the 4-(1-pyrenylvinyl)-N-butylpyridinium bromide; FITC is the fluorescein isothiocyanate; Cy3/3.5/5 are the cyanine 3/3.5/5; UP is the upconverting phosphor; TFP is the butterfly-shaped conjugated oligoelectrolyte; respectively.

7.2.2 Electrochemical Biosensors

Although GO is usually insulating with a large band gap, the CRGO (depending on the extent of reduction) can be as conducting as graphene and exhibit excellent electromechanical properties, as addressed in Chap. 3. Not only the holes of RGO

Table 7.1 GO as FRET acceptor in biosensing

Analyte	Biomolecular probe	Donor, E_{xc}/E_m (nm)	MQE (%)	Mechanism	LOD
ssDNA; Thrombin	ssDNA; Aptamer	FAM, 480/525	~96	π–π interactions	~2 nM
α-fetoprotein	Antibody	QD, 340/507	~70	Sandwich immunocomplex	~0.15 ng/mL
dsDNA	Charged dye	PNPB, 238/595	~90	Electrostatic interaction; π–π stacking	~1 nM
ssDNA	Molecular beacon	FAM, 480/520	~99	Hydrophobic and π–π interactions	~2 nM
ssDNAs	DNA probes	FAM, 494/526; Cy5, 643/666; ROX, 587/609	~97	Electrostatic interaction; π–π stacking	~100 pM
ssDNA	Molecular beacon	QD, 330/587	~98	π–π stacking; Hydrogen bonding	~12 nM
Thrombin	Aptamer	FAM, 470/517	~87	π–π stacking	31.3 nM
Cyclin A$_2$	Peptide	FITC, 495/520	~99	Electrostatic interaction; π–π stacking	0.5 nM
Helicase	ssDNA–GO complex	FAM, 480/520	~92	π–π stacking	N/A
ATP	Aptamer	FAM, 480/520	~100	π–π stacking	N/A
ATP	Molecular aptamer beacon	FAM, 480/520	~98	π–π stacking	~2 μM
ssDNA; Thrombin; Cysteine	Versatile molecular beacon	FITC, 480/520; Cy3, 520/565; Cy5, 630/665	~95	π–π stacking	~1 nM; ~5 nM; ~60 nM
Glucose	Concanavalin A	UP, 980/547	~81	π–π stacking	~25 nM
Thrombin	Peptide	FITC, 480/520	~90	π–π stacking	~2 nM
dsDNA	DNA probe	FAM, 480/520	~95	π–π stacking	~14 nM
Ochratoxin A	Aptamer	FAM, 492/520	~80	π–π stacking	~22 nM
ssDNA	DNA probe	SYBR green I, 490/529	~96	π–π stacking	~0.31 nM
Heparin	TFP	TFP, 380/529	~90	Electrostatic interaction; π–π stacking	0.046 U/mL
Metalloproteinase 2	Peptide	FITC, 470/516	~90	Electrostatic interaction; π–π stacking	50 pM
ssDNA	DNA probe	Cy3.5, 400/596	~75	π–π stacking	N/A
Enterovirus 71; Coxsackievirus B3	Antibody	QD 350/525; QD 350/605	~71 ~83	π–π stacking; Hydrogen bonding	0.42 ng/mL; 0.39 ng/mL

Reprinted with permission from Ref. [11]. Copyright (2012) Wiley-VCH Verlag GmbH & Co. KGaA, Weinheim

facilitate the transport of electrolyte ions, but also the functional groups accelerate the redox reactions. Particularly, RGO is capable of facilitating the direct electron transfer of enzymes and proteins at a GO-based electrode, which is promising for biosensors [28].

7.2.2.1 Enzyme Biosensing

RGO presents excellent performance for direct electrochemistry of GOD and can be used as GOD biosensors. Zhou et al. [16] demonstrated that RGO-based biosensor exhibits substantially enhanced amperometric signals for sensing glucose with a wide linear response range from 0.01 to 10 mM, a high sensitivity of 20.21 μA mM^{-1} cm^{-2} and a low detection limit of 2.00 μM (S/N = 3). The linear range for glucose detection is wider than that on the electrodes with other carbon materials, such as carbon nanotubes and nanofibers [15]. The response at the GOD/RGO/glassy carbon (GC) electrode to glucose is very fast (9 $\pm$ 1 s to steady-state response) and highly stable (91 % signal retention for 5 h). Qiu et al. [29] reported a glucose biosensor based on the electrodes consisting of homogeneous chitosan–ferrocene (CS–Fc)/GO/GOD nanocomposites. The uniformly dispersed GO within the CS matrix significantly improves the stability of GO and make it carrying a positive charge, which can further immobilize the GOD with higher loading. As a consequence, biosensors based on the CS–Fc/GO/GOD films show fast response, excellent reproducibility, high stability. Ping et al. [30] suggested that a disposable biosensor based on electrochemically reduced GO (ERGO) and GOD shows better analytical performance for the glucose detection compared with the biosensor based on ionic liquid doped screen-printed electrode. Using the ERGO based biosensor, the linear range for the detection of glucose is from 5.0 μM to 12.0 mM with a detection limit of 1.0 μM.

Furthermore, the complexes of RGO and metal nanoparticles have also been developed for glucose biosensing. For example, the RGO–PAMAM–silver nanoparticle (RGO–PAMAM–Ag) nanocomposite is a promising candidate material for high-performance glucose biosensors [31]. The glucose biosensor based on GOD electrode modified with RGO–PAMAM–Ag nanocomposites displays good performance, including high sensitivity (75.72 μA mM^{-1} cm^{-2}), low detection limit (4.5 μM), and an acceptable linear range from 0.032 to 1.89 mM. It also prevents the interference of some interfering species usually coexisting with glucose in the human blood at the work potential of –0.25 V [31]. Yang et al. [32] also confirmed the good performance of RGO-metal nanoparticle composites for enzymatic biosensing. They found that the ERGO–AuPd nanocomposites show excellent biocompatibility, enhanced electron transfer kinetics and large electroactive surface area, which are highly sensitive and stable towards oxygen reduction. Table 7.2 summarizes the performance of RGO-based materials for GOD biosensing.

In addition to the GOD biosensors, RGO-based electrodes have been employed to electrochemically biosense other enzymes with promising sensitivity and stability, such as horseradish peroxidase (HRP), alcohol dehydrogenase (ADH),

Table 7.2 The RGO-based electrodes for GOD biosensing

Electrode	Sensitivity (μA mM^{-1} cm^{-2})	Linear range (mM)	LOD	References
GOD/RGO/GC	20.21	0.01–10.0	2.0	[16]
GOD/CS–Fc/GO	10.0	0.02–6.8	7.6	[29]
GOD/E-RGO	22.8	0.005–12.0	1.0	[30]
GOD/RGO–PAMAM–Ag	75.7	0.032–1.89	4.5	[31]
GOD/RGO–AuPd	267	0.5–3.5	6.9	[32]

organophosphorus hydrolase (OPH), microperoxidase-11, tyrosinase, acetylcholinesterase, catalase, urease, and organophosphate pesticides (OPs) [33].

7.2.2.2 DNA Biosensing

Generally, electrochemical DNA sensors offer fast response speed, high sensitivity and selectivity, low cost for detecting selected DNA sequences, DNA hybridization and DNA damage, and specific analytes. The main mechanism of electrochemical DNA biosensors is the formation of DNA duplex from the probe/target DNAs, which involves the immobilization of ssDNA (the probe/target DNA) onto the electrode surface [33]. Among all kinds of electrochemical DNA sensors, the simplest one is based on the direct oxidation of DNA [16, 34, 35].

In 2009, Zhou et al. [16] designed an electrochemical DNA sensor with different electrodes to detect free bases of DNA, i.e., guanine (G), adenine (A), thymine (T), and cytosine (C), including the glassy carbon (GC), graphite/GC, and RGO/GC, as shown in Fig. 7.5. The current signals of the four free bases of DNA are all separated efficiently on the RGO/GC electrode, as illustrated in different colors in Fig. 7.5c. This indicates simultaneous detection of the four free bases, while neither graphite nor glassy carbon can. The main factors for simultaneous detection of the four free bases are the antifouling properties and the high electron transfer kinetics for bases oxidation on RGO/GC electrode, resulting from high density of edge-plane-like defective sites and oxygen-containing functional groups of the RGO, which provide many active sites and benefit acceleration of electron transfer between the electrode and species in solution. Moreover, the RGO/GC electrode is also able to efficiently separate all four DNA bases in both ssDNA and dsDNA at physiological pH without the need of a prehydrolysis step, as shown in Fig. 7.5e, f. This can be attributed to the unique physical and chemical properties of the RGO, such as the thickness of single sheet, high conductivity, large surface area, antifouling properties, and high electron transfer kinetics. In addition, these features also allows the RGO/GC electrode to detect a single-nucleotide polymorphism (SNP) site for short oligomers with a particular sequence without any hybridization or labeling processes.

By directly assembling captured ssDNA onto a RGO-modified electrode via the π–π stacking interaction between the RGO and the ssDNA bases, Lin et al. [36]

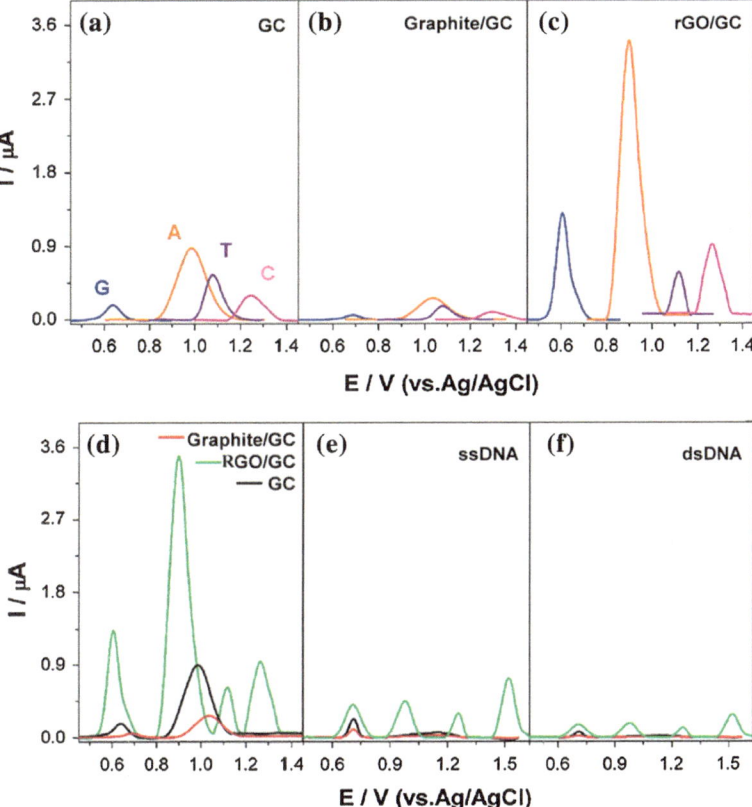

Fig. 7.5 Differential pulse voltammograms (DPVs) at the GC (**a**), graphite/GC (**b**) and RGO/GC (**c**) electrodes for free bases of G (*blue*), A (*orange*), T (*violet*), and C (*magenta*), respectively; DPVs for a mixture of G, A, T, and C (**d**), ssDNA (**e**), dsDNA (**f**) at graphite/GC (*red*), RGO/GC (*green*), and GC electrodes (*black*), respectively. Reprinted with permission from Ref. [16]. Copyright (2009) American Chemical Society

fabricated an electrochemical DNA biosensor for sensitive and selective detection of DNA. The biosensor is in a sandwich assay format mixed with the target DNA sequence and oligonucleotide probe-labeled AuNPs, followed by the AuNP-catalyzed silver deposition. Due to the high DNA loading ability of the RGO and the distinct signal amplification by AuNP-catalyzed silver staining, the fabricated biosensor exhibits good analytical performance with a wide detection linear range from 200 pM to 500 nM and a low detection limit of 72 pM.

7.2.2.3 Other Molecule Biosensing

In addition to enzyme and DNA, GO-based electrodes also exhibit electrochemical sensing capabilities for other biomolecules, such as hydrazine, paracetamol, and DA [33]. Based on a chemically reduced poly sodium styrenesulfonate (PSS)/RGO nanocomposite film, Wang et al. [37] proposed an ultrasensitive biosensor for the determination of hydrazine. Taking advantage of the high specific surface area and higher electrical conductivity of the composite materials, the fabricated biosensor achieves a calibration curve for hydrazine with a linear range of 3.0–300 μmol L^{-1} and a detection limit as low as 1 μmol L^{-1}. Similarly, Kang et al. [38] constructed an GO-based electrochemical sensor for the sensitive detection of paracetamol and found excellent performance for detecting paracetamol with a detection limit of 3.2×10^{-8} M, a reproducibility of 5.2 % relative standard deviation, and a satisfied recovery from 96.4 to 103.3 %. Besides, Zhou et al. [16] designed an electrochemical sensor with RGO/GC electrodes to detect DA, which exhibits more favorable electron transfer kinetics than the graphite/GC and GC electrodes.

7.2.3 Enzyme Inhibitors

In addition, GO can also serve as an enzyme inhibitor. De et al. [39] indicated that GO exhibits the highest inhibition dose response for α–chymotrypsin (ChT) inhibition compared with all other artificial inhibitors reported. Besides, it was found that GO can be cytotoxic from investigation of the effect of fetal bovine serum (FBS) on the cytotoxicity of GO [40]. Figure 7.6 illustrates the interaction between the GO and the α–chymotrypsin. The cytotoxicity of GO nanosheets occurs as a result of physical damage to the cell membrane, which is induced by direct interactions between the cell membrane and the GO nanosheets. However, the cytotoxicity of GO nanosheets occurs mostly during the initial contact stage of GO and cells and is greatly mitigated at 10 % FBS.

Lin et al. [41] studied the activity and inhibition of T4 Polynucleotide Kinase (PNK) by GO. Due to the extraordinary fluorescence quenching ability of GO, sensitive detection of PNK activity and inhibitor screening were achieved. Then Jin et al. [42] investigated the role of GO for selective modulating the enzyme activity and thermostability by analyzing the interactions between serine proteases and GO functionalized with different amine-terminated PEG. Three well-characterized serine proteases were studied, including the trypsin, chymotrypsin, and proteinase K. It was found that without PEGylation, as-made GO significantly inhibited enzyme activity likely due to the nonspecific binding of enzyme proteins on GO via hydrophobic interactions. In this case, active center of the enzyme is blocked or denatured, leading to the dramatically decreased enzyme activity. However, when the GO was functionalized with PEG, it could selectively improve trypsin activity and thermostability with 60–70 % retained activity at 80 °C. In contrast, the PEGylated GO shows barely any effect on chymotrypsin or proteinase K.

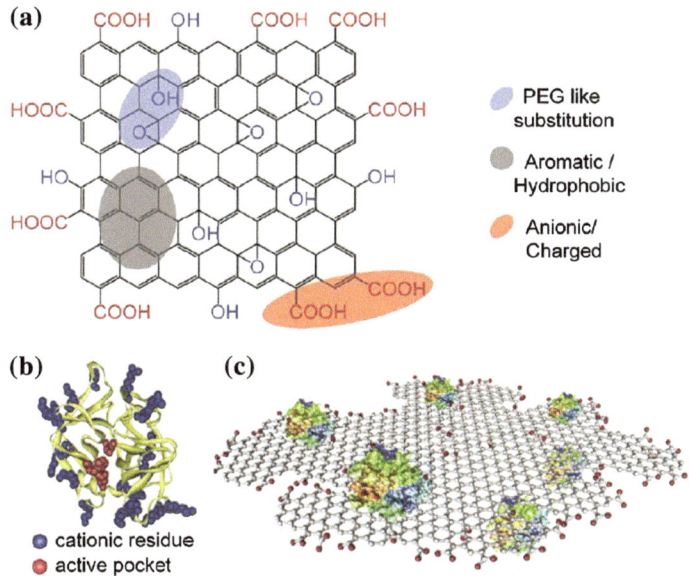

Fig. 7.6 Schematic structures of GO (**a**), α-chymotrypsin (**b**), and GO and protein complexation (**c**). Reprinted with permission from Ref. [39]. Copyright (2011) American Chemical Society

7.2.4 Drug Delivery

Another exciting application of GO in biotechnology is the drug delivery in living cells [1]. It was found that the PEG-modified GO can act as a platform for the delivery of water-soluble cancer drugs. Sun et al. [43] studied the possibility of nanoscale GO (NGO) for drug delivery. They first checked the toxicity of NGO-PEG by incubating Raji cells in various concentrations of NGO-PEG for 72 h. They found that only for extremely high NGO-PEG concentrations (>100 mg/L), the cell viability becomes slightly reduced. Then doxorubicin (DOX), a widely used cancer drug, was loaded onto ~2 mg/L of NGO-PEG (a concentration far below the toxic level) by simple physisorption via π–π stacking. Moreover, the drug release from NGO-PEG is pH-dependent, as shown in Fig. 7.7a. In an acidic solution of pH 5.5, ~40 % of DOX loaded on NGO-PEG is released over 1 day. However, at a pH of 7.4, the release rate is reduced, ~15 % over 2 days.

   Similarly, Liu et al. [44] investigated the effect of NGO on delivering SN38. The detailed procedures are described as follows. First, NGO is functionalized with PEG to render high solubility in aqueous solutions and stability in physiological solutions like serum. Then, a water-insoluble aromatic molecule, SN38, is attached to NGO-PEG. Finally, the NGO-PEG-SN38 complex exhibit high potency with IC50 values of 6 nM for HCT-116 human colon cancer cells, which

(a)   **(b)**

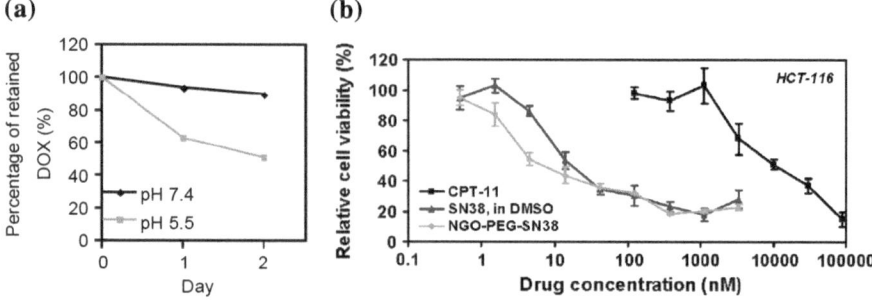

Fig. 7.7 a Retained DOX as a function of time on NGO-PEG in buffers at pH 5.5 and 7.4. Reprinted with permission from Ref. [43]. Copyright (2008) Tsinghua Press and Springer-Verlag. **b** Relative cell viability (versus untreated control) data of HCT-116 cells incubated with CPT-11, SN38, and NGO-PEG-SN38 at different concentrations for 72 h. Reprinted with permission from Ref. [44]. Copyright (2008) American Chemical Society

is 1,000-fold more potent than camptothecin (CPT-11), and comparable to that of free SN38 dissolved in DMSO [43], as displayed in Fig. 7.7b.

In addition, to enhance the loading efficiency and targeting ability of anticancer drugs, NGO can be covalently modified with folic acid (FA) [45]. Zhang et al. [45] proposed a strategy to controllable loading of two anticancer drugs (DOX and CPT-11) onto the FA-conjugated NGO (FA-NGO). As a result, the FA-NGO loaded with the two anticancer drugs shows specific targeting to MCF-7 human breast cancer cells and remarkably high cytotoxicity compared to unmodified NGO loaded with DOX or irinotecan.

References

1. Wang, Y., Li, Z., Wang, J., Li, J., Lin, Y.: Trends Biotechnol. **29**, 205–212 (2011)
2. Zhang, J., Zhang, F., Yang, H., Huang, X., Liu, H., Zhang, J., Guo, S.: Langmuir **26**, 6083–6085 (2010)
3. Kotchey, G.P., Allen, B.L., Vedala, H., Yanamala, N., Kapralov, A.A., Tyurina, Y.Y., Klein-Seetharaman, J., Kagan, V.E., Star, A.: ACS Nano **5**, 2098–2108 (2011)
4. Chen, C., Xie, Q., Yang, D., Xiao, H., Fu, Y., Tan, Y., Yao, S.: RSC Adv. **3**, 4473–4491 (2013)
5. Wang, Z., Zhou, X., Zhang, J., Boey, F., Zhang, H.: J. Phys. Chem. C **113**, 14071–14075 (2009)
6. Liu, Y., Yu, D., Zeng, C., Miao, Z., Dai, L.: Langmuir **26**, 6158–6160 (2010)
7. Liu, J., Li, Y., Li, Y., Li, J., Deng, Z.: J. Mater. Chem. **20**, 900–906 (2010)
8. Xu, Y., Wu, Q., Sun, Y., Bai, H., Shi, G.: ACS Nano **4**, 7358–7362 (2010)
9. Wu, M., Kempaiah, R., Huang, P.J.J., Maheshwari, V., Liu, J.: Langmuir **27**, 2731–2738 (2011)
10. Zhang, M., Yin, B.C., Tan, W., Ye, B.C.: Biosens. Bioelectron. **26**, 3260–3265 (2011)
11. Morales-Narváez, E., Merkoçi, A.: Adv. Mater. **24**, 3298–3308 (2012)
12. Lu, C.H., Yang, H.H., Zhu, C.L., Chen, X., Chen, G.N.: Angew. Chem. **121**, 4879–4881 (2009)
13. He, S., Song, B., Li, D., Zhu, C., Qi, W., Wen, Y., Wang, L., Song, S., Fang, H., Fan, C.: Adv. Funct. Mater. **20**, 453–459 (2010)

14. Dong, H., Gao, W., Yan, F., Ji, H., Ju, H.: Anal. Chem. **82**, 5511–5517 (2010)
15. Shao, Y., Wang, J., Wu, H., Liu, J., Aksay, I.A., Lin, Y.: Electroanalysis **22**, 1027–1036 (2010)
16. Zhou, M., Zhai, Y., Dong, S.: Anal. Chem. **81**, 5603–5613 (2009)
17. Tang, L., Wang, Y., Li, Y., Feng, H., Lu, J., Li, J.: Adv. Funct. Mater. **19**, 2782–2789 (2009)
18. Xu, L.Q., Yang, W.J., Neoh, K.-G., Kang, E.T., Fu, G.D.: Macromolecules **43**, 8336–8339 (2010)
19. Bulinski, J.C.: Int. Rev. Cytol. **103**, 281–302 (1986)
20. Han, T.H., Lee, W.J., Lee, D.H., Kim, J.E., Choi, E.Y., Kim, S.O.: Adv. Mater. **22**, 2060–2064 (2010)
21. Crawford, R.L.: Lignin Biodegradation and Transformation. Wiley, New York (1981)
22. Yang, Q., Pan, X., Huang, F., Li, K.: J. Phys. Chem. C **114**, 3811–3816 (2010)
23. Jung, J.H., Cheon, D.S., Liu, F., Lee, K.B., Seo, T.S.: Angew. Chem. Int. Ed. **49**, 5708–5711 (2010)
24. Liu, F., Choi, J.Y., Seo, T.S.: Biosens. Bioelectron. **25**, 2361–2365 (2010)
25. Song, E., Cheng, D., Song, Y., Jiang, M., Yu, J., Wang, Y.: Biosens. Bioelectron. **47**, 41445–41450 (2013)
26. Liu, M., Zhao, H., Quan, X., Chen, S., Fan, X.: Chem. Commun. **46**, 7909–7911 (2010)
27. Wang, X., Wang, C., Qu, K., Song, Y., Ren, J., Miyoshi, D., Sugimoto, N., Qu, X.: Adv. Funct. Mater. **20**, 3967–3971 (2010)
28. Zuo, X., He, S., Li, D., Peng, C., Huang, Q., Song, S., Fan, C.: Langmuir **26**, 1936–1939 (2009)
29. Qiu, J.D., Huang, J., Liang, R.P.: Sens. Actuators, B **160**, 287–294 (2011)
30. Ping, J., Wang, Y., Fan, K., Wu, J., Ying, Y.: Biosens. Bioelectron. **28**, 204–209 (2011)
31. Luo, Z., Yuwen, L., Han, Y., Tian, J., Zhu, X., Weng, L., Wang, L.: Biosens. Bioelectron. **36**, 179–185 (2012)
32. Yang, J., Deng, S., Lei, J., Ju, H., Gunasekaran, S.: Biosens. Bioelectron. **29**, 159–166 (2011)
33. Chen, D., Feng, H., Li, J.: Chem. Rev. **112**, 6027–6053 (2012)
34. Niwa, O., Jia, J., Sato, Y., Kato, D., Kurita, R., Maruyama, K., Suzuki, K., Hirono, S.: J. Am. Chem. Soc. **128**, 7144–7145 (2006)
35. Drummond, T.G., Hill, M.G., Barton, J.K.: Nat. Biotechnol. **21**, 1192–1199 (2003)
36. Lin, L., Liu, Y., Tang, L., Li, J.: Analyst **136**, 4732–4737 (2011)
37. Wang, C., Zhang, L., Guo, Z., Xu, J., Wang, H., Zhai, K., Zhuo, X.: Microchim. Acta **169**, 1–6 (2010)
38. Kang, X., Wang, J., Wu, H., Liu, J., Aksay, I.A., Lin, Y.: Talanta **81**, 754–759 (2010)
39. De, M., Chou, S.S., Dravid, V.P.: J. Am. Chem. Soc. **133**, 17524–17527 (2011)
40. Hu, W., Peng, C., Lv, M., Li, X., Zhang, Y., Chen, N., Fan, C., Huang, Q.: ACS Nano **5**, 3693–3700 (2011)
41. Lin, L., Liu, Y., Zhao, X., Li, : J. Anal. Chem. **83**, 8396–8402 (2011)
42. Jin, L., Yang, K., Yao, K., Zhang, S., Tao, H., Lee, S.T., Liu, Z., Peng, R.: ACS Nano **6**, 4864–4875 (2012)
43. Sun, X., Liu, Z., Welsher, K., Robinson, J.T., Goodwin, A., Zaric, S., Dai, H.: Nano Res. **1**, 203–212 (2008)
44. Liu, Z., Robinson, J.T., Sun, X., Dai, H.: J. Am. Chem. Soc. **130**, 10876–10877 (2008)
45. Zhang, L., Xia, J., Zhao, Q., Liu, L., Zhang, Z.: Small **6**, 537–544 (2010)

Chapter 8
Conclusion and Outlook

As a superstar nanomaterial, GO has gained tremendous attentions in recent years and holds great promise for many technological applications. So far, various methods have been developed to fabricate GO samples and to characterize their atomic structures. Meanwhile, GO is an important raw material to mass-produce graphene via reduction. The reduced graphene oxide, namely, RGO, is also attractive from materials science point of view.

Due to the abundant oxygen-containing functional groups and controllable ratio of sp^2 to sp^3 hybridized carbon atoms, GO possesses tunable electronic and optical properties, which further lead to exciting applications in several fields. For example, the controllable band gap makes GO good electrical and optical devices, such as electrical sensors, field-emission devices, and photovoltaic devices. Intrinsic fluorescence in wide regions also enables GO for usage of optical detecting and sensing.

The existence of functional groups on GO and RGO makes them easily to composite with other nanomaterials such as nanoparticles, carbon nanotubes, and conducting polymers. GO/RGO and their composites show great potentials in the energy storage/conversion and environmental protection technologies, e.g., photocatalyst for water splitting under solar irradiation, hydrogen storage medium, electrodes for various lithium batteries and supercapacitors, and removal of pollutants in air and water. Moreover, the functional groups make GO hydrophilic and dissolvable, and further noncovalently interact with biomolecules. As a result, GO is very useful in biotechnology to detect biomolecules with high sensitivity and selectivity and to deliver drugs.

Despite the comprehensive research efforts and advances described in this book, there are still many critical issues of GO itself to be answered. For example, the atomic structure of GO is still under debate. Many experimental characterizations demonstrate that the functional groups are randomly distributed in the graphene basal plane, while theoretical simulations predict that GO structures with hydroxyl and epoxy groups orderly aligned together along the armchair direction show preferable thermodynamic stability. Investigation of the structural evolution during oxidation/reduction processes is also of key importance to understand the

© The Author(s) 2015
J. Zhao et al., *Graphene Oxide: Physics and Applications*,
SpringerBriefs in Physics, DOI 10.1007/978-3-662-44829-8_8

structural details of GO, such as formation of holes, non-hexagonal carbon rings, and variation of functional groups. On the other hand, only some fundamental reactions for the oxidation and reduction mechanism of GO have been studied. The detailed mechanisms, especially for the multi-step and complex reactions, still need to be elucidated.

Although numerous promising applications have been proposed, the commercial and industrial applications of GO are still at the starting stage with many remaining difficulties and challenges. For GO itself, the chemical and photocatalytic mechanisms in generation and storage of hydrogen, oxidation or reduction of harmful gases and metal ions, and degradation of organic species are partly unclear. The role of surface chemistry and morphology control of GO on the electrochemical performance in lithium batteries and supercapacitors also needs systematical explorations. In principles, there are nearly infinite combinations of GO/RGO with other functional nanomaterials, making the GO/RGO based composites a largely unexplored area with many opportunities. Understanding the interaction mechanism between GO/RGO with the nanomaterials in those nanoscale hybrids from the atomistic point of view is imperative. Deeper insights into the intrinsic synergistic effects of GO/RGO composites in improving the overall performance of energy storage and conversion, environmental protection and biotechnology are desirable.

Considering the outstanding performance of the GO-based materials in many aspects, we anticipate more researches from fundamental science as well as more R&D projects of GO from the industries.